Synthesis Lectures on Computer Vision

Series Editors

Gerard Medioni, University of Southern California, Los Angeles, CA, USA

Sven Dickinson, Department of Computer Science, University of Toronto, Toronto, ON, Canada

This series publishes on topics pertaining to computer vision and pattern recognition. The scope follows the purview of premier computer science conferences, and includes the science of scene reconstruction, event detection, video tracking, object recognition, 3D pose estimation, learning, indexing, motion estimation, and image restoration. As a scientific discipline, computer vision is concerned with the theory behind artificial systems that extract information from images. The image data can take many forms, such as video sequences, views from multiple cameras, or multi-dimensional data from a medical scanner. As a technological discipline, computer vision seeks to apply its theories and models for the construction of computer vision systems, such as those in self-driving cars/navigation systems, medical image analysis, and industrial robots.

Lei Huang

Normalization Techniques in Deep Learning

 Springer

Lei Huang
Beihang University
Beijing, China

ISSN 2153-1056   ISSN 2153-1064 (electronic)
Synthesis Lectures on Computer Vision
ISBN 978-3-031-14597-1   ISBN 978-3-031-14595-7 (eBook)
https://doi.org/10.1007/978-3-031-14595-7

This Springer imprint is published by the registered company Springer Nature Switzerland AG
The registered company address is: Gewerbestrasse 11, 6330 Cham, Switzerland

Preface

I focused my research on normalization techniques in deep learning since 2015, when the milestone technique—batch normalization (BN)—was published. I witness the research progresses of normalization techniques, including the analyses for understanding the mechanism behind the design of corresponding algorithms and the application for particular tasks. Despite the abundance and ever more important roles of normalization techniques, there is an absence of a unifying lens with which to describe, compare, and analyze them. In addition, our understanding of theoretical foundations of these methods for their success remains elusive.

This book provides a research landscape for normalization techniques, covering methods, analyses, and applications. It can provide valuable guidelines for selecting normalization techniques to use in training DNNs. With the help of these guidelines, it will be possible for students/researchers to design new normalization methods tailored to specific tasks or improve the trade-off between efficiency and performance. As the key components in DNNs, normalization techniques are links that connect the theory and application of deep learning. We thus believe that these techniques will continue to have a profound impact on the rapidly growing field of deep learning, and we hope that this book will aid researchers in building a comprehensive landscape for their implementation.

This book is based on our survey paper [1], but with significant extents and updates, including but not limited to the details of techniques and the recent research progresses. The target audiences of this book are graduate students, researchers, and practitioners who have been working on development of novel deep learning algorithms and/or their application to solve practical problems in computer vision and machine learning tasks. For intermediate and advanced level researchers, this book presents theoretical analysis of normalization methods and the mathematical tools used to develop new normalization methods.

Beijing, China

Lei Huang

February 2022

Reference

1. Huang, L., J. Qin, Y. Zhou, F. Zhu, L. Liu, and L. Shao (2020). Normalization techniques in training DNNs: Methodology, analysis and application. *CoRR abs/2009.12836*.

Acknowledgements

I would like to express my sincere thanks to all those who worked with me on this book. Some parts of this book are based on our previous published or pre-printed papers [1–9], and I am grateful to my co-authors in the work of normalization technique in deep learning: Xianglong Liu, Adams Wei Yu, Bo Li, Jia Deng, Dawei Yang, Bo Lang, Dacheng Tao, Yi Zhou, Jie Qin, Fan Zhu, Li Liu, Ling Shao, Lei Zhao, Diwen Wan, Yang Liu, Zehuan Yuan. In addition, this work is supported by the National Natural Science Foundation of China (No.62106012). I would also thank all the editors, reviewers, and staff who helped with the publication of the book. Finally, I thank my family for their wholehearted support.

Beijing, China   Lei Huang
February 2022

References

1. Huang, L., X. Liu, Y. Liu, B. Lang, and D. Tao (2017). Centered weight normalization in accelerating training of deep neural networks. In *ICCV*.
2. Huang, L., X. Liu, B. Lang, and B. Li (2017). Projection based weight normalization for deep neural networks. *arXiv preprint* arXiv:1710.02338.
3. Huang, L., D. Yang, B. Lang, and J. Deng (2018). Decorrelated batch normalization. In *CVPR*.
4. Huang, L., X. Liu, B. Lang, A. W. Yu, Y. Wang, and B. Li (2018). Orthogonal weight normalization: Solution to optimization over multiple dependent stiefel manifolds in deep neural networks. In *AAAI*.
5. Huang, L., Y. Zhou, F. Zhu, L. Liu, and L. Shao (2019). Iterative normalization: Beyond standardization towards efficient whitening. In *CVPR*.
6. Huang, L., L. Zhao, Y. Zhou, F. Zhu, L. Liu, and L. Shao (2020). An investigation into the stochasticity of batch whitening. In *CVPR*.
7. Huang, L., L. Liu, F. Zhu, D. Wan, Z. Yuan, B. Li, and L. Shao (2020). Controllable orthogonalization in training DNNs. In *CVPR*.

8. Huang, L., J. Qin, L. Liu, F. Zhu, and L. Shao (2020). Layer-wise conditioning analysis in exploring the learning dynamics of DNNs. In *ECCV*.
9. Huang, L., Y. Zhou, L. Liu, F. Zhu, and L. Shao (2021). Group whitening: Balancing learning efficiency and representational capacity. In *CVPR*.

Contents

About the Author

Lei Huang is currently an Associate Professor in Institute of Artificial Intelligence of Beihang University, China. He received his B.Sc. and Ph.D. degrees in 2010 and 2018, respectively, at the School of Computer Science and Engineering, Beihang University. From 2015 to 2016, he visited the Vision and Learning Lab, University of Michigan, Ann Arbor, USA. During 2018 to 2020, he was a Research Scientist in Inception Institute of Artificial Intelligence (IIAI), UAE. His current research mainly focuses on normalization techniques (involving methods, theories, and applications) in training DNNs. He also has wide interests in deep learning theory (representation & optimization) and computer vision tasks. He serves as a Reviewer for the top conferences and journals such as CVPR, ICCV, ECCV, NeurIPS, AAAI, JMLR, IJCV, TPAMI, etc.

Introduction

Deep neural networks (DNNs) have been extensively used across a broad range of applications, including computer vision (CV), natural language processing (NLP), speech and audio processing, robotics, bioinformatics, etc. [1]. They are typically composed of stacked layers/modules, the transformation between which consists of a linear mapping with learnable parameters and a nonlinear activation function [2]. While their deep and complex structure provides them powerful representation capacity and appealing advantages in learning feature hierarchies, it also makes their training difficult [3, 4]. One notorious problem in training DNNs is the so-called activations (and gradients) vanishing or exploding, which is mainly caused by the compounded linear or nonlinear transformation in DNNs. Here, we provide an illustration in Fig. 1.1. When activations vanishing occurs, a DNN may map all the inputs into the same representation, which make the inputs indistinguishable and thus the learning will be difficult (even impossible). When activations exploding occurs, a DNN will exaggerate the disturbance of the input, which may cause catastrophe in learning dynamics. Similarly, vanishing or exploding gradients will also impair the learning, as illustrated in [4].

In fact, the success of DNNs heavily depends on breakthroughs in training techniques [5–8], especially on controlling the distribution of activations by design, which has been witnessed by the history of deep learning [1]. For example, Hinton and Salakhutdinov proposed layer-wise initialization that pioneers the research on good initialization methods for linear layers, aiming to implicitly designing well shaped activations/gradients during initialization. This makes training deep models possible. One milestone technique in addressing the training issues of DNNs was batch normalization (BN) [8], which explicitly standardizes the activations of intermediate DNN layers within a mini-batch of data. BN improves DNNs' training stability, optimization efficiency and generalization ability. It is a basic component in most state-of-the-art architectures [9–16], and has successfully proliferated throughout various areas of deep learning [17–19]. By now, BN is used by default in most deep learning models, both in research (more than 34,000 citations on Google scholar) and real-world settings [20]. Further, a significant number of other normalization techniques

© The Author(s), under exclusive license to Springer Nature Switzerland AG 2022
L. Huang, *Normalization Techniques in Deep Learning*, Synthesis Lectures on Computer
Vision, https://doi.org/10.1007/978-3-031-14595-7_1

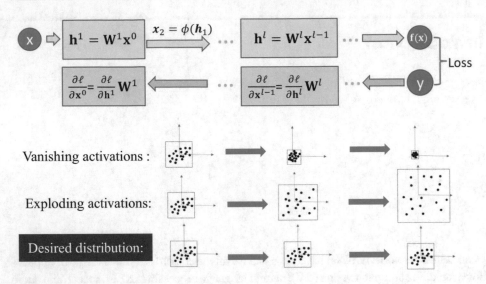

Fig. 1.1 Illustration of the vanishing exploding activations in a DNN

have been proposed to address the training issues in particular contexts, further evolving the DNN architectures and their applications [21–25]. For example, layer normalization (LN) [21] is an essential module in Transformer [26], which has advanced the state-of-the-art architectures for NLP [26–29], while spectral normalization [24] is a basic component in the discriminator of generative adversarial networks (GANs) [24, 30, 31]. Importantly, the ability of most normalization techniques to stabilize and accelerate training has helped to simplify the process of designing network architectures—training is no longer the main concern, enabling more focus to be given to developing components that can effectively encode prior/domain knowledge into the architectures.

However, despite the abundance and ever more important roles of normalization techniques, we note that there is an absence of a unifying lens with which to describe, compare and analyze them [32]. It is essential to provide guidelines for elaborating, understanding and applying normalization methods. This book provides a review and commentary on normalization techniques in the context of training DNNs. We attempt to provide answers for the following questions:

(1) What are the main motivations behind different normalization methods in DNNs, and how can we present a taxonomy for understanding the similarities and differences between a wide variety of approaches?
(2) How can we reduce the gap between the empirical success of normalization techniques and our theoretical understanding of them?
(3) What recent advances have been made in designing/tailoring normalization techniques for different tasks, and what are the main insights behind them?

We answer the first question by providing a unified picture of the main motivations behind different normalization methods, from the perspective of optimization (Chap. 2). We show that most normalization methods are essentially designed to satisfy nearly equal statistical distributions of layer input/output-gradients across different layers during training, in order to avoid the ill-conditioned landscape of optimization. Based on this, we provide a comprehensive review of the normalization methods, including a general view of normalizing activations (Chap. 3), normalizing activations by population statistics (Sect. 3.1, Chap. 6), normalizing activations as functions (Chap. 4 and 5), normalizing weights (Chap. 7) and normalizing gradients (Chaps. 8). Specifically, we decompose the most representative normalizing-activations-as-functions framework into three components: the normalization area partitioning (NAP), normalization operation (NOP) and normalization representation recovery (NRR). We unify most normalizing-activations-as-function methods into this framework, and provide insights for designing new normalization methods.

To answer the second question, we discuss the recent progress in our theoretical understanding of BN in Chap. 9. It is difficult to fully analyze the inner workings of BN in a unified framework, but our review ultimately provides clear guidelines for understanding why BN stabilizes and accelerates training, and further improves generalization, through a scale-invariant analysis, condition analysis and stochasticity analysis, respectively.

We answer the third question in Chap. 10 by providing a review of the applications of normalization for particular tasks, and illustrating how normalization methods can be used to solve key issues. To be specific, we mainly review the applications of normalization in domain adaptation, style transfer, training GANs and efficient deep models. We show that the normalization methods can be used to 'edit' the statistical properties of layer activations. These statistical properties, when designed well, can represent the style information for a particular image or the domain-specific information for a distribution of a set of images. This characteristic of normalization methods has been thoroughly exploited in CV tasks and potentially beyond them.

We conclude the book with additional thoughts about certain open questions in the research of normalization techniques.

1.1   Denotations and Definitions

In this book, we use a lowercase letter $x \in \mathbb{R}$ to denote a scalar, boldface lowercase letter $\mathbf{x} \in \mathbb{R}^d$ for a vector, boldface uppercase letter for a matrix $X \in \mathbb{R}^{d \times m}$, and boldface sans-serif notation for a tensor X, where $\mathbb{R}$ is the set of real-valued numbers, and d, m are positive integers. Note that a tensor is a more general entity. Scalars, vectors and matrices can be viewed as 0th-order, 1st-order and 2nd-order tensors. Here, X denotes a tensor with an order larger than 2. We will provide a more precise definition in the later sections. We follow matrix notation where the vector is in column form, except that the derivative is a row vector.

1.1.1 Optimization Objective

Consider a true data distribution $p_*(\mathbf{x}, \mathbf{y}) = p(\mathbf{x})p(\mathbf{y}|\mathbf{x})$ and the sampled training sets $\mathbb{D} \sim$ $p_*(\mathbf{x}, \mathbf{y})$ of size N: $\mathbb{D} = \{(\mathbf{x}^{(i)}, \mathbf{y}^{(i)})\}_{i=1}^{N}$. We focus on a supervised learning task aiming to learn the conditional distribution $p(\mathbf{y}|\mathbf{x})$ using the model $q(\mathbf{y}|\mathbf{x})$, where $q(\mathbf{y}|\mathbf{x})$ is represented as a function $\mathbf{f}_\theta(\mathbf{x})$ parameterized by θ. Training the model can be viewed as tuning the parameters to minimize the discrepancy between the desired output $\mathbf{y}$ and the predicted output $\mathbf{f}(\mathbf{x}; \theta)$. This discrepancy is usually described by a loss function $\ell(\mathbf{y}, \mathbf{f}(\mathbf{x}; \theta))$ for each sample pair $(\mathbf{x}, \mathbf{y})$. The empirical risk, averaged over the sample loss in training sets $\mathbb{D}$, is defined as:

$$\mathcal{L}(\theta) = \frac{1}{N} \sum_{i=1}^{N} (\ell(\mathbf{y}^{(i)}, \mathbf{f}_\theta(\mathbf{x}^{(i)}))). \tag{1.1}$$

This book mainly focuses on discussing the empirical risk from the perspective of optimization. We do not explicitly analyze the risk under the true data distribution $\mathcal{L}^*(\theta) = \mathbb{E}_{(\mathbf{x},\mathbf{y}) \sim p_*(\mathbf{x},\mathbf{y})}(\ell(\mathbf{y}, \mathbf{f}_\theta(\mathbf{x})))$ from the perspective of generalization.

1.1.2 Neural Networks

The function $\mathbf{f}(\mathbf{x}; \theta)$ adopted by neural networks usually consists of stacked layers. For a multilayer perceptron (MLP), $\mathbf{f}_\theta(\mathbf{x})$ can be represented as a layer-wise linear and nonlinear transformation (Fig. 1.2), as follows:

$$\mathbf{h}^l = \mathbf{W}^l \mathbf{x}^{l-1}, \tag{1.2}$$

$$\mathbf{x}^l = \phi(\mathbf{h}^l), \quad l = 1, \ldots, L, \tag{1.3}$$

where $\mathbf{x}^0 = \mathbf{x}$, $\mathbf{W}^l \in \mathbb{R}^{d_l \times d_{l-1}}$ and d_l indicates the number of neurons in the l-th layer. The learnable parameters $\theta = \{\mathbf{W}^l, l = 1, \ldots, L\}$. Typically, $\mathbf{h}^l$ and $\mathbf{x}^l$ are referred to as the pre-activation and activation, respectively, but in this book, we refer to both as activations for simplicity, if we do not explicitly differentiate them. We also set $\mathbf{x}^L = \mathbf{h}^L$ as the output of the network $\mathbf{f}_\theta(\mathbf{x})$ to simplify denotations.

Convolutional Layer: The convolutional layer parameterized by weights $\mathbf{W} \in \mathbb{R}^{d_l \times d_{l-1} \times F_h \times F_w}$, where F_h and F_w are the height and width of the filter, takes feature maps (activations) $\mathbf{X} \in \mathbb{R}^{d_{l-1} \times h \times w}$ as input, where h and w are the height and width of the feature maps, respectively. We denote the set of spatial locations as Δ and the set of spatial offsets as Ω. For each output feature map k and its spatial location $\delta \in \Delta$, the convolutional layer computes the pre-activation $\{H_{k,\delta}\}$ as: $H_{k,\delta} = \sum_{i=1}^{d_{l-1}} \sum_{\tau \in \Omega} W_{k,i,\tau} X_{i,\delta+\tau} = \langle \mathbf{w}_k, \mathbf{x}_\delta \rangle$. Therefore, the convolution operation is a linear (dot) transformation. Here, $\mathbf{w}_k \in \mathbb{R}^{d_{l-1} \cdot F_h \cdot F_w}$ can eventually be viewed as an unrolled filter produced by $\mathbf{W}$.

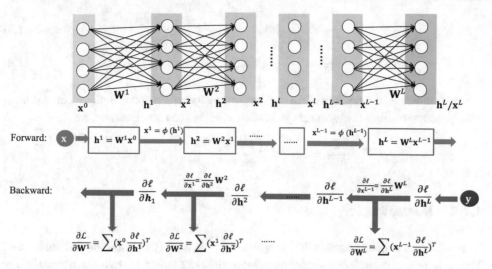

Fig. 1.2 Illustration of the representation, forward calculation and back-propagation of a MLP model

1.1.3 Training DNNs

From an optimization perspective, we aim to minimize the empirical risk $\mathcal{L}$, as:

$$\theta^* = \arg\min_\theta \mathcal{L}(\theta). \tag{1.4}$$

In general, the gradient descent (GD) update is used to minimize $\mathcal{L}$, seeking to iteratively reduce the loss as:

$$\theta_{t+1} = \theta_t - \eta \frac{\partial \mathcal{L}}{\partial \theta}, \tag{1.5}$$

where η is the learning rate. For large-scale learning, stochastic gradient descent (SGD) is extensively used to approximate the gradients $\frac{\partial \mathcal{L}}{\partial \theta}$ with the gradient sampled from a mini-batch data with size of m: $\mathcal{B} = (\mathbf{x}^{(i)}, \mathbf{y}^{(i)})_{i=1}^m$, as:

$$\frac{\partial \mathcal{L}}{\partial \theta} = \frac{1}{m} \sum_{i=1}^m \frac{\partial \ell^{(i)}}{\partial \theta}, \tag{1.6}$$

where $\frac{\partial \ell^{(i)}}{\partial \theta}$ is the abbreviation of the gradient *w.r.t.* the i-th sample: $\frac{\partial \ell(\mathbf{y}^{(i)}, \mathbf{f}_\theta(\mathbf{x}^{(i)}))}{\partial \theta}$. One essential step is to calculate the gradients. This can be done by back-propagation (Fig. 1.2) for calculating $\frac{\partial \ell}{\partial \mathbf{x}^{l-1}}$ and $\frac{\partial \ell}{\partial \mathbf{h}^{l-1}}$:

$$\frac{\partial \ell}{\partial \mathbf{x}^{l-1}} = \frac{\partial \ell}{\partial \mathbf{h}^l} W^l, \tag{1.7}$$

$$\frac{\partial \ell}{\partial \mathbf{h}^{l-1}} = \frac{\partial \ell}{\partial \mathbf{x}^{l-1}} \odot \phi'(\mathbf{h}^{l-1}), \ l = L, \ldots, 2, \tag{1.8}$$

where ϕ' indicates the gradient of the non-linear transformation and $\odot$ indicates the element-wise product. Further, the gradients *w.r.t.* each layer W^l can be calculated as:

$$\frac{\partial \mathcal{L}}{\partial W^l} = \mathbb{E}_{\mathbb{D}}[(\mathbf{x}^{l-1} \frac{\partial \ell}{\partial \mathbf{h}^l})^T], \quad l = L, \ldots, 1. \tag{1.9}$$

1.1.4 Normalization

Normalization is widely used in data-preprocessing [9, 33, 34], data mining and other areas. The definition of normalization may vary among different topics. In statistics, normalization can have a range of meanings [35]. In the simplest cases, normalization of ratings means adjusting values measured on different scales to a common scale. In more complicated cases, normalization may refer to more sophisticated adjustments where the intention is to bring the entire probability distributions of adjusted values into alignment. In image processing, normalization is a process that changes the range of pixel intensity values. It is sometimes called contrast stretching or histogram stretching [36]. Applications include photographs with poor contrast due to glare.

In this book, we define normalization as a general transformation, which ensures that the transformed data has certain statistical properties. To be more specific, we provide the following formal definition.

Definition 1.1 Normalization: Given a set of data $\mathbb{D} = \{\mathbf{x}^{(i)}\}_{i=1}^N$, the normalization operation is a function $\Phi : \mathbf{x} \longmapsto \hat{\mathbf{x}}$, which ensures that the transformed data $\widehat{\mathbb{D}} = \{\hat{\mathbf{x}}^{(i)}\}_{i=1}^N$ has certain statistical properties.

We consider five main normalization operations (Fig. 1.3) in this book: centering, scaling, decorrelating, standardizing and whitening [37].

Centering formulates the transformation as:

$$\hat{\mathbf{x}} = \Phi_C(\mathbf{x}) = \mathbf{x} - \mu, \tag{1.10}$$

where $\mu = \mathbb{E}_{\mathbb{D}}(\mathbf{x})$ is the mean of $\mathbf{x}$. This ensures that the normalized output $\hat{\mathbf{x}}$ has a zero-mean property, which can be represented as: $\mathbb{E}_{\widehat{\mathbb{D}}}(\mathbf{x}) = \mathbf{0}$.

Scaling formulates the transformation as:

$$\hat{\mathbf{x}} = \Phi_{SC}(\mathbf{x}) = \Lambda^{-\frac{1}{2}}\mathbf{x}. \tag{1.11}$$

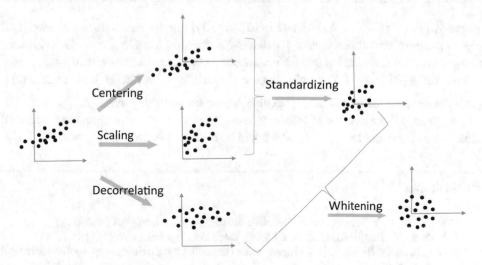

Fig. 1.3 Illustration of normalization operations discussed in this book

Here, $\Lambda = \mathrm{diag}(\sigma_1^2, \dots, \sigma_d^2)$, where σ_j^2 is the mean square over data samples for the j-th dimension: $\sigma_j^2 = \mathbb{E}_{\mathbb{D}}(\mathbf{x}_j^2)$. Scaling ensures that the normalized output $\hat{\mathbf{x}}$ has a unit-variance property, which can be represented as: $\mathbb{E}_{\widehat{\mathbb{D}}}(\hat{\mathbf{x}}_j^2) = 1$ for all $j = 1, \dots, d$.

Decorrelating formulates the transformation as:

$$\hat{\mathbf{x}} = \Phi_D(\mathbf{x}) = \boldsymbol{D}^T \mathbf{x}, \tag{1.12}$$

where $\boldsymbol{D} = [\mathbf{d}_1, \dots, \mathbf{d}_d]$ are the eigenvectors of Σ from eig decomposition and $\Sigma = \mathbb{E}_{\mathbb{D}}(\mathbf{x}\mathbf{x}^T)$ is the covariance matrix of $\mathbf{x}$. Decorrelating ensures that the correlation between different dimensions of the normalized output $\hat{\mathbf{x}}$ is zero (*i.e.*, the covariance matrix $\mathbb{E}_{\widehat{\mathbb{D}}}(\hat{\mathbf{x}}\hat{\mathbf{x}}^T)$ is a diagonal matrix).

Standardizing is a composition operation that combines centering and scaling, as:

$$\hat{\mathbf{x}} = \Phi_{ST}(\mathbf{x}) = \Lambda^{-\frac{1}{2}}(\mathbf{x} - \mu). \tag{1.13}$$

Standardizing ensures that the normalized output $\hat{\mathbf{x}}$ has zero-mean and unit-variance properties.

Whitening formulates the transformation as[1]:

$$\hat{\mathbf{x}} = \Phi_W(\mathbf{x}) = \widetilde{\Lambda}^{-\frac{1}{2}} \boldsymbol{D}^T \mathbf{x}, \tag{1.14}$$

---

[1] Whitening usually requires the input to be centered [37, 38], which means it also includes the centering operation. In this book, we unify the operation as whitening regardless of whether it includes centering or not.

where $\widetilde{\Lambda}_d = \mathrm{diag}(\lambda_1, \ldots, \lambda_d)$ and $\boldsymbol{D} = [\mathbf{d}_1, \ldots, \mathbf{d}_d]$ are the eigenvalues and associated eigenvectors of covariance matrix Σ from eig decomposition $\boldsymbol{D}\widetilde{\Lambda}_d\boldsymbol{D}^T = \Sigma$. Whitening ensures that the normalized output $\hat{\mathbf{x}}$ has a spherical Gaussian distribution, which can be represented as: $\mathbb{E}_{\mathbb{D}}(\hat{\mathbf{x}}\hat{\mathbf{x}}^T) = \mathbf{I}$. The whitening transformation, defined in Eq. 1.14, is called principal components analysis (PCA) whitening, where the whitening matrix $\Sigma_{PCA}^{-\frac{1}{2}} = \widetilde{\Lambda}_d^{-\frac{1}{2}}\boldsymbol{D}$. There are an infinite number of whitening matrices since a whitened input stays whitened after an arbitrary rotation [37, 39], which will be discussed in Sect. 4.2 for details.

References

1. Goodfellow, I., Y. Bengio, and A. Courville (2016). *Deep Learning*. MIT Press.
2. LeCun, Y., Y. Bengio, and G. Hinton (2015). Deep learning. *nature 521*(7553), 436–444.
3. Glorot, X. and Y. Bengio (2010). Understanding the difficulty of training deep feedforward neural networks. In *AISTATS*.
4. Pascanu, R., T. Mikolov, and Y. Bengio (2013). On the difficulty of training recurrent neural networks. In *ICML*.
5. Hinton, G. E. and R. R. Salakhutdinov (2006). Reducing the dimensionality of data with neural networks. *science 313*(5786), 504–507.
6. Nair, V. and G. E. Hinton (2010). Rectified linear units improve restricted boltzmann machines. In *ICML*.
7. Kingma, D. P. and J. Ba (2015). Adam: A method for stochastic optimization. In *ICLR*.
8. Ioffe, S. and C. Szegedy (2015). Batch normalization: Accelerating deep network training by reducing internal covariate shift. In *ICML*.
9. He, K., X. Zhang, S. Ren, and J. Sun (2016a). Deep residual learning for image recognition. In *CVPR*.
10. Szegedy, C., W. Liu, Y. Jia, P. Sermanet, S. Reed, D. Anguelov, D. Erhan, V. Vanhoucke, and A. Rabinovich (2015). Going deeper with convolutions. In *CVPR*.
11. Zagoruyko, S. and N. Komodakis (2016). Wide residual networks. In *BMVC*.
12. Szegedy, C., V. Vanhoucke, S. Ioffe, J. Shlens, and Z. Wojna (2016). Rethinking the inception architecture for computer vision. In *CVPR*.
13. He, K., X. Zhang, S. Ren, and J. Sun (2016b). Identity mappings in deep residual networks. In *ECCV*.
14. Qi, C. R., H. Su, K. Mo, and L. J. Guibas (2017). Pointnet: Deep learning on point sets for 3d classification and segmentation. In *CVPR*.
15. Huang, G., Z. Liu, and K. Q. Weinberger (2017). Densely connected convolutional networks. In *CVPR*.
16. Xie, S., R. B. Girshick, P. Dollár, Z. Tu, and K. He (2017). Aggregated residual transformations for deep neural networks. In *CVPR*.
17. Russakovsky, O., J. Deng, H. Su, J. Krause, S. Satheesh, S. Ma, Z. Huang, A. Karpathy, A. Khosla, M. Bernstein, et al. (2015). Imagenet large scale visual recognition challenge. *International journal of computer vision 115*(3), 211–252.
18. Lin, T.-Y., M. Maire, S. Belongie, J. Hays, P. Perona, D. Ramanan, P. Dollár, and C. L. Zitnick (2014). Microsoft coco: Common objects in context. In *ECCV*, pp. 740–755.
19. Chang, A. X., T. Funkhouser, L. Guibas, P. Hanrahan, Q. Huang, Z. Li, S. Savarese, M. Savva, S. Song, H. Su, et al. (2015). Shapenet: An information-rich 3d model repository. *arXiv preprint* arXiv:1512.03012.

20. Santurkar, S., D. Tsipras, A. Ilyas, and A. Madry (2018). How does batch normalization help optimization? In *NeurIPS*.
21. Ba, L. J., R. Kiros, and G. E. Hinton (2016). Layer normalization. *arXiv preprint* arXiv:1607.06450.
22. Salimans, T. and D. P. Kingma (2016). Weight normalization: A simple reparameterization to accelerate training of deep neural networks. In *NeurIPS*.
23. Wu, Y. and K. He (2018). Group normalization. In *ECCV*.
24. Miyato, T., T. Kataoka, M. Koyama, and Y. Yoshida (2018). Spectral normalization for generative adversarial networks. In *ICLR*.
25. Huang, L., L. Liu, F. Zhu, D. Wan, Z. Yuan, B. Li, and L. Shao (2020). Controllable orthogonalization in training DNNs. In *CVPR*.
26. Vaswani, A., N. Shazeer, N. Parmar, J. Uszkoreit, L. Jones, A. N. Gomez, L. Kaiser, and I. Polosukhin (2017). Attention is all you need. In *NeurIPS*.
27. Yu, A. W., D. Dohan, M.-T. Luong, R. Zhao, K. Chen, M. Norouzi, and Q. V. Le (2018). Qanet: Combining local convolution with global self-attention for reading comprehension. In *ICLR*.
28. Xu, J., X. Sun, Z. Zhang, G. Zhao, and J. Lin (2019). Understanding and improving layer normalization. In *NeurIPS*.
29. Xiong, R., Y. Yang, D. He, K. Zheng, S. Zheng, C. Xing, H. Zhang, Y. Lan, L. Wang, and T.-Y. Liu (2020). On layer normalization in the transformer architecture. In *ICML*.
30. Kurach, K., M. Lučić, X. Zhai, M. Michalski, and S. Gelly (2019). A large-scale study on regularization and normalization in GANs. In *ICML*.
31. Brock, A., J. Donahue, and K. Simonyan (2019). Large scale GAN training for high fidelity natural image synthesis. In *ICLR*.
32. Huang, L., J. Qin, Y. Zhou, F. Zhu, L. Liu, and L. Shao (2020). Normalization techniques in training DNNs: Methodology, analysis and application. *CoRR abs/2009.12836*.
33. LeCun, Y., L. Bottou, G. B. Orr, and K.-R. Muller (1998). Efficient backprop. In *Neural Networks: Tricks of the Trade*.
34. Krizhevsky, A. (2009). Learning multiple layers of features from tiny images. Technical report.
35. Dodge, Y. (2003). *The Oxford Dictionary of Statistical Terms,*. Oxford University Press.
36. Gonzalez, R. C. and R. E. Woods (2008). *Digital image processing*. Prentice Hall.
37. Kessy, A., A. Lewin, and K. Strimmer (2018). Optimal whitening and decorrelation. *The American Statistician 72*(4), 309–314.
38. Huang, L., D. Yang, B. Lang, and J. Deng (2018). Decorrelated batch normalization. In *CVPR*.
39. Huang, L., L. Zhao, Y. Zhou, F. Zhu, L. Liu, and L. Shao (2020). An investigation into the stochasticity of batch whitening. In *CVPR*.

Motivation and Overview of Normalization in DNNs

Input normalization is extensively used in machine learning models, and is advocated in many textbooks or courses for that the improved performance of the learner can be obtained. Intuitively, normalizing an input removes the difference in magnitude between different features of examples, which benefits learning. This characteristic is important for non-parametric models, for example the K-nearest neighbor (KNN) classifier, in which we need to calculate the distance or similarity between the examples. If certain dimensions of features have large range in magnitude, the distance/similarity between examples will be dominated by these dimensions, which will impair the performance of the learner. Besides, normalizing an input can improve the optimization efficiency for parametric models. There exist theoretical advantages to normalization for linear models, as we will illustrate.

2.1   Theory of Normalizing Input

Let us consider a linear regression model with a scalar output $f_{\mathbf{w}}(\mathbf{x}) = \mathbf{w}^T \mathbf{x}$ where $\mathbf{x}, \mathbf{w} \in \mathbb{R}^d$, and mean square error (MSE) loss $\ell = \frac{1}{2}(y - f_{\mathbf{w}}(\mathbf{x}))^2$. Given a set of data $\mathbb{D} = \{(\mathbf{x}^{(i)}, y^{(i)})\}_{i=1}^N$, the cost function can be calculated as

$$\mathcal{L}(\mathbf{w}) = \frac{1}{2N} \sum_{i=1}^N (y^{(i)} - f_{\mathbf{w}}(\mathbf{x}^{(i)}))^2. \qquad (2.1)$$

Note that the cost function of Eq. 2.1 is quadratic in $\mathbf{w}$. It can be rewritten as

$$\mathcal{L}(\mathbf{w}) = \frac{1}{2}(\mathbf{w}^T \Sigma \mathbf{w} - 2\mathbf{a}^T \mathbf{w} + b), \qquad (2.2)$$

where $\Sigma \in \mathbb{R}^{d \times d}$ is the covariance matrix of the input, calculated by $\Sigma = \frac{1}{N} \sum_{n=1}^N \mathbf{x}^{(i)}(\mathbf{x}^{(i)})^T$; the d-dimensional vector $\mathbf{a} = \frac{1}{N} \sum_{i=1}^N y^{(i)}\mathbf{x}^{(i)}$; the constant

© The Author(s), under exclusive license to Springer Nature Switzerland AG 2022
L. Huang, *Normalization Techniques in Deep Learning*, Synthesis Lectures on Computer Vision, https://doi.org/10.1007/978-3-031-14595-7_2

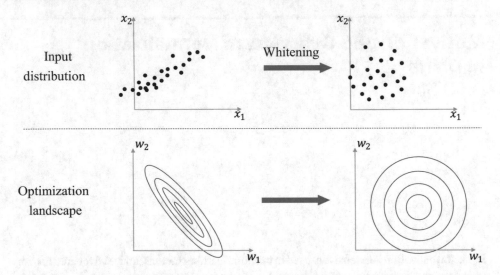

Fig. 2.1 Theoretical motivation of normalization input from the perspective of optimization. For a linear regression model with MSE loss, its optimization landscape is solely controlled by the distribution of input data

$b = \frac{1}{N} \sum_{i=1}^{N} (y^{(i)})^2$. We note that the gradient of the cost function is given by $\frac{\partial \mathcal{L}}{\partial \mathbf{w}} = \mathbf{w}^T \Sigma - \mathbf{a}^T$, while the Hessian matrix of second derivatives is $H = \Sigma$. The landscape of $\mathcal{L}(\mathbf{w})$ is therefore determined by the covariance matrix of the input. If the input is whitened by Eq. 1.14, the surface of $\mathcal{L}(\mathbf{w})$ will be isometric (Fig. 2.1). Under this case, the GD can converge within only one iteration. There also exists theory [1, 2] analytically showing that the learning dynamics for such a quadratic surface are fully controlled by the spectrum of the Hessian matrix. Here, we illustrate it in details.

We now consider conducting a coordinate transformation $\mathbf{v} = Trans(\mathbf{w})$ such that $\mathcal{L}(\mathbf{v})$ has a decoupled form with respect to $\mathbf{v}$. Let $\mathbf{w}^*$ denote the solution space which minimize $\mathcal{L}(\mathbf{w})$. We have $\Sigma \mathbf{w}^* = \mathbf{a}$, based on that the minimal is obtained under the condition $\frac{\partial \mathcal{L}}{\partial \mathbf{w}} = \mathbf{0}$. Note that $\mathbf{w}^*$ reduces to a point and has a closed form $\mathbf{w}^* = \Sigma^{-1} \mathbf{a}$ if Σ is full rank. We now conduct a coordinate transformation $\mathbf{v}' = \mathbf{w} - \mathbf{w}^*$, which provides new coordinates centered at the solution point as:

$$\mathcal{L}(\mathbf{v}') = \frac{1}{2}(\mathbf{v}')^T \Sigma \mathbf{v}' + \mathcal{L}_0, \tag{2.3}$$

with $\mathcal{L}_0 = \mathcal{L}(\mathbf{w}^*)$. We can further decorrelate the surface of cost function by conducting a coordinate transformation $\mathbf{v} = D\mathbf{v}'$ as:

$$\mathcal{L}(\mathbf{v}) = \frac{1}{2}\mathbf{v}^T \widetilde{\Lambda} \mathbf{v} + \mathcal{L}_0, \tag{2.4}$$

where $\widetilde{\Lambda} = \mathrm{diag}(\lambda_1, \ldots, \lambda_d)$ and $D = [\mathbf{d}_1, \ldots, \mathbf{d}_d]$ are the eigenvalues and associated eigenvectors of covariance matrix Σ. The Hessian matrix is the diagonal matrix $\widetilde{\Lambda}$ in this new coordinate system with respect to $\mathbf{v}$, after the compounded transformation:

$$\mathbf{v} = D(\mathbf{w} - \mathbf{w}^*). \tag{2.5}$$

The weight update thus is formulated as:

$$\mathbf{v}(t+1) = \mathbf{v}(t) - \eta\widetilde{\Lambda}\mathbf{v}(t). \tag{2.6}$$

This can be viewed as a set of d decoupled equations. Given the initial state $\mathbf{v}(0)$, the evolution of each component of $\mathbf{v}$ is given by

$$\mathbf{v}_j(t) = (1 - \eta\lambda_j)^t \mathbf{v}_j(0), \, for \, j = 1, \ldots, d. \tag{2.7}$$

We note $\mathbf{v}_j = 0$ implies $\mathbf{w}_j = \mathbf{w}_j^*$, based on Eq. 2.5. We can see the convergence behaviors of $\mathbf{w}_j$ depends not only on the learning rate η, but also the corresponding eigenvalue λ_j of covariance matrix. We also note that $\mathbf{v}_j$ will not converge if $\lambda_j = 0$, since $\mathbf{v}_j = \mathbf{v}_j(0)$ during the training. Now we can break down into more cases, if $\lambda_j > 0$:

(1) $0 < \eta < \frac{2}{\lambda_j}$. $\mathbf{v}_j$ will converge exponentially to zero, with characteristic time $\tau_j = (\eta\lambda_j)^{-1}$ [1]. Furthermore, if $\frac{1}{\lambda_j} < \eta < \frac{2}{\lambda_j}$, it converges to the solution in an oscillatory way; while $0 < \eta < \frac{1}{\lambda_j}$, it converges monotonically. In the case of $\eta = \frac{1}{\lambda_j}$, the convergence for $\mathbf{v}_j$ will occur in only one iteration, which corresponds to Newton's method.

(2): $\eta = \frac{2}{\lambda_j}$. The $\mathbf{v}_j$ will oscillate between $\pm\mathbf{v}_j(0)$.

(3) $\eta > \frac{2}{\lambda_j}$. The $\mathbf{v}_j$ will diverge exponentially.

Let's consider all the eigenvalues together. Convergence requires $0 < \eta < \frac{2}{\lambda_j}$ for all $1 \le j \le d$, and η thus must be in the range $0 < \eta < \frac{2}{\lambda_{max}}$, where λ_{max} is the largest eigenvalue of Σ. The slowest time constant in this objective is corresponding to the direction of the lowest nonzero eigenvalue λ_{min}, with characteristic time $\tau_{max} = (\eta\lambda_{min})^{-1}$. The optimal learning rate should be $\eta = \frac{1}{\lambda_{max}}$, which thus leads to $\tau_{max} = \frac{\lambda_{max}}{\lambda_{min}}$. In optimization community, $\kappa = \frac{\lambda_{max}}{\lambda_{min}}$ is call the condition number of the corresponding curvature matrix, and it controls the number of iterations required for convergence (e.g., the lower bound of the iterations is κ [1–3]).

If all eigenvalues are equal, i.e., $\lambda_j = \lambda$ for all $1 < j < d$, the Hessian matrix is isometric diagonal. Convergence can be obtained in one iteration, with optimal learning rate $\eta = \frac{1}{\lambda}$. Indeed, this situation can be obtained by whitening the input through Eq. 1.14, as illustrated before. Therefore, normalizing the input can surely accelerate convergence during the optimization of linear models. Particularly, this behavior can be well characterized based on the spectral analysis of the covariance matrix of the input data.

Linear regression with multiple neuron. It is easy to extend the solution of linear regression from a scalar output to a vectorial output $\mathbf{f}_W(\mathbf{x}) = W^T\mathbf{x}$. In this case, the Hessian is represented as

$$H = \mathbb{E}_{\mathbb{D}}(\mathbf{x}\mathbf{x}^T) \otimes I, \tag{2.8}$$

where $\otimes$ indicates the Kronecker product [4] and I denotes the identity matrix. We can see the curvature of the cost function is still controlled by the covariance matrix of the input

Linear Classification. For the linear classification model with cross entropy loss, the Hessian is approximated by [5]:

$$H \approx \mathbb{E}_{\mathbb{D}}(\mathbf{x}\mathbf{x}^T) \otimes S. \tag{2.9}$$

$S \in \mathbb{R}^{c \times c}$ is defined by $S = \frac{1}{c}(I_c - \frac{1}{c}\mathbf{1}_c\mathbf{1}_c^T)$, where c is the number of classes and $\mathbf{1}_c \in \mathbb{R}^c$ denotes a vector in which all entries are 1. Equation 2.9 assumes the Hessian does not significantly change from the initial region to the optimal region [5].

2.2   Towards Normalizing Activations

These theoretical results for linear model do not applies to deep neural networks (DNNs) directly, because DNNs are non-linear models. Even though DNNs include layer-wise linear transformation, however the input $\mathbf{x}$ is only linearly connected to the first weight matrix W^1 (Fig. 1.2); but the optimization is over all the parameters $\theta = \{W^i, i = 1, \ldots, L\}$, not W^1 only. Despite of the non-linearity, we can exploit the layer-wise structure of DNNs. We can find that the layer-wise activation $\mathbf{x}^l$ is linearly connected by the following weight matrix W^{l+1}. This provides an intuition that normalizing activations can benefit the optimization, based on the theoretical results for linear models. We will illustrate the motivation from the perspectives of proximal back-propagation framework and Fisher information matrix (FIM) approximation.

2.2.1   Proximal Back-Propagation Framework

Carreira-Perpinan and Wang [6] proposed to use auxiliary coordinates to redefine the optimization object $\mathcal{L}(\theta)$ with equality constraints imposed on each neuron. They solved the constrained optimization by adding a quadratic penalty as:

$$\widetilde{\mathcal{L}}(\theta, \{\mathbf{x}^l\}_{l=1}^L) = \mathcal{L}(\mathbf{y}, \mathbf{f}^L(W^L, \mathbf{x}^{L-1})) + \sum_{l=1}^{L-1} \frac{\alpha}{2}\|\mathbf{x}^l - \mathbf{f}^l(W^l, \mathbf{x}^{l-1}))\|^2, \tag{2.10}$$

where $\mathbf{f}^l(\cdot, \cdot)$ is a function transformation with respect to the l-th layer. As shown in [6], the solution for minimizing $\widetilde{\mathcal{L}}(\theta, \{\mathbf{x}^l\}_{l=1}^L)$ converges to the solution for minimizing $\mathcal{L}(\theta)$ as $\alpha \to \infty$, under mild conditions. Furthermore, the proximal propagation [7] and the following back-matching propagation [8] reformulate each sub-problem independently with a backward order, minimizing each layer object $\mathcal{L}^l(W^l, \mathbf{x}^{l-1}; \tilde{\mathbf{x}}^l)$, given the target signal $\tilde{\mathbf{x}}^l$ from the upper layer, as follows:

$$\begin{cases} \mathcal{L}(\mathbf{y}, \mathbf{f}^L(W^L, \mathbf{x}^{L-1})), & for \ l = L \\ \frac{1}{2}\|\tilde{\mathbf{x}}^l - \mathbf{f}^l(W^l, \mathbf{x}^{l-1})\|^2, & for \ l = L-1, \dots, 1. \end{cases} \tag{2.11}$$

It has been shown that the produced W^l using gradient update w.r.t. $\mathcal{L}(\theta)$ equals to the W^l produced by the back-matching propagation (Procedure 1 in [8]) with one-step gradient update w.r.t. Eq. 2.11, given an appropriate step size. Note that the target signal $\tilde{\mathbf{x}}^l$ is obtained by back-propagation, which means the loss $\mathcal{L}(\theta)$ would be smaller if $\mathbf{f}^l(W^l, \mathbf{x}^{l-1})$ is more close to $\tilde{\mathbf{x}}^l$. The loss $\mathcal{L}(\theta)$ will be reduced more efficiently, if the sub-optimization problems in Eq. 2.11 are well-conditioned. Please refer to [7, 8] for more details. If we view the auxiliary variable as the pre-activation in a specific layer, the sub-optimization problem in each layer is formulated as:

$$\begin{cases} \mathcal{L}(\mathbf{y}, W^L \mathbf{x}^{L-1}), & for \ l = L \\ \frac{1}{2}\|\tilde{\mathbf{x}}^l - W^l \mathbf{x}^{l-1}\|^2, & for \ l = L-1, \dots, 1. \end{cases} \tag{2.12}$$

It is clear that the sub-optimization problems with respect to W^l are actually linear classification (for $l = L$) or regression (for $l = 1, \dots, L-1$) models. Their conditioning analysis is thoroughly characterized in Sect. 2.1.

This connection suggests that the quality (conditioning) of the full optimization problem $\mathcal{L}(\theta)$ is well correlated to its sub-optimization problems shown in Eq. 2.12, whose local curvature matrix can be well explored. Thus, normalizing the activations can benefit optimization.

2.2.2 K-FAC Approximation

Another theoretic base of normalizing the activations benefiting optimization is from the perspective of approximating the Fisher information matrix (FIM). The FIM can be used to approximately describe the optimization landscapes. It is computationally expensive to calculate the FIM of DNNs, due to the large parameters. There exist methods to approximate the FIM.

One successful example is approximating the FIM of DNNs using the Kronecker product (K-FAC) [9–12]. In the K-FAC approach, there are two assumptions: (1) weight-gradients in different layers are assumed to be uncorrelated; (2) the input and output-gradient in each layer are approximated as independent. Thus, the full FIM can be represented as a block diagonal matrix, $\mathbf{F} = diag(F^1, \dots, F^L)$, where F^l is the sub-FIM (the FIM with respect to the parameters in a certain layer) and computed as:

$$F^l \approx \mathbb{E}_{\mathbf{x} \sim p(\mathbf{x})}[\mathbf{x}^{l-1}(\mathbf{x}^{l-1})^T] \otimes \mathbb{E}_{(\mathbf{x},\mathbf{y}) \sim p(\mathbf{x})q(\mathbf{y}|\mathbf{x})}\left[\frac{\partial \ell}{\partial \mathbf{h}^l}^T \frac{\partial \ell}{\partial \mathbf{h}^l}\right]. \tag{2.13}$$

For the details of derivation, please refer to [9]. Note that Huang et al. [13], Martens [9], Martens and Grosse [14] have provided empirical evidence to support their effectiveness in approximating the full FIM with block diagonal sub-FIMs.

We can see that each sub-FIM is controlled by its covariance matrix of layer input and the covariance matrix of output-gradients. Therefore, the optimization landscape of the optimization problems can be approximately controlled by the covariance matrix of the input and the output-gradient.

2.2.3   Highlights of Motivation

We denote the covariance matrix of the layer input as $\Sigma_{\mathbf{x}}^l = \mathbb{E}_{p(x)q(y|x)}[\mathbf{x}^{l-1}(\mathbf{x}^{l-1})^T]$ and the covariance matrix of the layer output-gradient as $\Sigma_{\nabla\mathbf{h}}^l = \mathbb{E}_{q(\mathbf{y}|\mathbf{x})}[\frac{\partial\ell}{\partial\mathbf{h}^l}^T\frac{\partial\ell}{\partial\mathbf{h}^l}]$. Based on the K-FAC, it is clear that the conditioning of the FIM can be improved, if:

- *Criteria 1*: The statistics of the layer input (e.g., $\Sigma_{\mathbf{x}}$) and output-gradient (e.g., $\Sigma_{\nabla\mathbf{h}}$) across different layers are equal.
- *Criteria 2*: $\Sigma_{\mathbf{x}}$ and $\Sigma_{\nabla\mathbf{h}}$ are well conditioned.

A variety of techniques for training DNNs have been designed to satisfy *Criteria 1* and/or *Criteria 2*, essentially. For example, the weight initialization techniques aim to satisfy *Criteria 1*, obtaining nearly equal variances for layer input/output-gradients across different layers [15–19] by designing initial weight matrices. However, the equal-variance property across layers can be broken down and is not necessarily sustained throughout training, due to the update of weight matrices. From this perspective, it is important to normalize the activations in order to produce better-conditioned optimization landscapes, similar to the benefits of normalizing the input.

Normalizing activations is more challenging than normalizing the input with a fixed distribution, since the distribution of layer activations $\mathbf{x}^l$ varies during training. Besides, DNNs are usually optimized over stochastic or mini-batch gradients, rather than the full gradient, which requires more efficient statistical estimations for activations. This book discusses three types of normalization methods for improving the performance of DNN training:

(1) Normalizing the activations directly to (approximately) satisfy Criteria 1 and/or Criteria 2. Generally speaking, there are two strategies for normalizing the activations of DNNs. One is to normalize the activations using the population statistics estimated over the distribution of activations [20–22]. The other strategy is to normalize the activations as a function transformation, which requires backpropagation through this transformation [23–25].

(2) Normalizing the weights with a constrained distribution, such that the activations/output-gradients (Eqs. 1.2 and 1.7) can be implicitly normalized. This normalization strategy is

inspired by weight initialization methods, but extends them towards satisfying the desired property during training [26–29].

(3) Normalizing gradients to exploit the curvature information for GD/SGD, even though the optimization landscape is ill-conditioned [30, 31]. This involves performing normalization solely on the gradients, which may effectively remove the negative effects of the ill-conditioned landscape caused by the diversity of magnitude in gradients from different layers (i.e., Criteria 1 is not well satisfied) [15].

References

1. LeCun, Y., I. Kanter, and S. A. Solla (1990). Second order properties of error surfaces. In *NeurIPS*.
2. LeCun, Y., L. Bottou, G. B. Orr, and K.-R. Muller (1998). Efficient backprop. In *Neural Networks: Tricks of the Trade*.
3. Bottou, L., F. E. Curtis, and J. Nocedal (2018). Optimization methods for large-scale machine learning. *Siam Review 60*(2), 223–311.
4. Grosse, R. B. and J. Martens (2016). A kronecker-factored approximate fisher matrix for convolution layers. In *ICML*, Volume 48, pp. 573–582.
5. Wiesler, S. and H. Ney (2011). A convergence analysis of log-linear training. In *NeurIPS*, pp. 657–665.
6. Carreira-Perpinan, M. and W. Wang (2014). Distributed optimization of deeply nested systems. In *AISTATS*.
7. Frerix, T., T. Möllenhoff, M. Möller, and D. Cremers (2018). Proximal backpropagation. In *ICLR*.
8. Zhang, H., W. Chen, and T.-Y. Liu (2018). On the local Hessian in back-propagation. In *NeurIPS*.
9. Martens, J. and R. Grosse (2015). Optimizing neural networks with kronecker-factored approximate curvature. In *ICML*.
10. Ba, J., R. Grosse, and J. Martens (2017). Distributed second-order optimization using kronecker-factored approximations. In *ICLR*.
11. Sun, K. and F. Nielsen (2017). Relative Fisher information and natural gradient for learning large modular models. In *ICML*.
12. Bernacchia, A., M. Lengyel, and G. Hennequin (2018). Exact natural gradient in deep linear networks and its application to the nonlinear case. In *NeurIPS*.
13. Martens, J. (2014). New perspectives on the natural gradient method. *arXiv preprint* arXiv:1412.1193.
14. Huang, L., J. Qin, L. Liu, F. Zhu, and L. Shao (2020). Layer-wise conditioning analysis in exploring the learning dynamics of DNNs. In *ECCV*.
15. Glorot, X. and Y. Bengio (2010). Understanding the difficulty of training deep feedforward neural networks. In *AISTATS*.
16. He, K., X. Zhang, S. Ren, and J. Sun (2015). Delving deep into rectifiers: Surpassing human-level performance on imagenet classification. In *ICCV*.
17. Saxe, A. M., J. L. McClelland, and S. Ganguli (2014). Exact solutions to the nonlinear dynamics of learning in deep linear neural networks. In *ICLR*.
18. Mishkin, D. and J. Matas (2016). All you need is a good init. In *ICLR*.
19. Sokol, P. A. and I. M. Park (2020). Information geometry of orthogonal initializations and training. In *ICLR*.

20. Montavon, G. and K.-R. Müller (2012). *Deep Boltzmann Machines and the Centering Trick*, Volume 7700.
21. Wiesler, S., A. Richard, R. Schlüter, and H. Ney (2014). Mean-normalized stochastic gradient for large-scale deep learning. In *ICASSP*.
22. Desjardins, G., K. Simonyan, R. Pascanu, and k. kavukcuoglu (2015). Natural neural networks. In *NeurIPS*.
23. Ioffe, S. and C. Szegedy (2015). Batch normalization: Accelerating deep network training by reducing internal covariate shift. In *ICML*.
24. Ba, L. J., R. Kiros, and G. E. Hinton (2016). Layer normalization. *arXiv preprint* arXiv:1607.06450.
25. Huang, L., D. Yang, B. Lang, and J. Deng (2018). Decorrelated batch normalization. In *CVPR*.
26. Salimans, T. and D. P. Kingma (2016). Weight normalization: A simple reparameterization to accelerate training of deep neural networks. In *NeurIPS*.
27. Huang, L., X. Liu, Y. Liu, B. Lang, and D. Tao (2017). Centered weight normalization in accelerating training of deep neural networks. In *ICCV*.
28. Huang, L., X. Liu, B. Lang, A. W. Yu, Y. Wang, and B. Li (2018). Orthogonal weight normalization: Solution to optimization over multiple dependent stiefel manifolds in deep neural networks. In *AAAI*.
29. Miyato, T., T. Kataoka, M. Koyama, and Y. Yoshida (2018). Spectral normalization for generative adversarial networks. In *ICLR*.
30. Yu, A. W., L. Huang, Q. Lin, R. Salakhutdinov, and J. Carbonell (2017). Block-normalized gradient method: An empirical study for training deep neural network. *arXiv preprint* arXiv:1707.04822.
31. You, Y., I. Gitman, and B. Ginsburg (2017). Large batch training of convolutional networks. *arXiv preprint* arXiv:1708.03888.

A General View of Normalizing Activations

3

As illustrated before, normalizing input, as a pre-processing, is usually performed over the full dataset, for improving the conditioning of optimization problem. It is straightforward to normalize the activations in deep neural networks over the full dataset using the population statistics, which is the main thoughts of normalization developed in machine learning communities. Besides, There are another line of work for normalizing the internal representation of deep neural network in computer vision communities to adjust the contrast information for images. In this chapter, we will first introduce the preliminary work of normalizing activations of DNNs, prior to the milestone normalization technique—batch normalization (BN) [1]. We then illustrate the algorithm of BN and how it is developed by exploiting the merits of the previous methods.

3.1   Normalizing Activations by Population Statistics

In this section, we will discuss the methods that normalize activations using the population statistics estimated over their distribution. This normalization strategy views the population statistics as constant during backpropagation. To simplify the notation, we remove the layer index l of activations $\mathbf{x}^l$ in the subsequent sections, unless otherwise stated.

Montavon and Müller [2] proposed to center the activations (hidden units) in a Boltzmann machine to improve the conditioning of the optimization problems, based on the insight that centered inputs improve the conditioning [3–5]. Specifically, given the activation in a certain layer $\mathbf{x}$, they perform the normalization as:

$$\hat{\mathbf{x}} = \mathbf{x} - \hat{\mu}, \tag{3.1}$$

where $\hat{\mu}$ is the mean of activations over the training dataset $\hat{\mu} = \mathbb{E}_{\mathbf{x} \sim \mathbb{D}}(\mathbf{x})$. Note that $\hat{\mu}$ indicates the population statistics that need to be estimated during training, and is considered

© The Author(s), under exclusive license to Springer Nature Switzerland AG 2022
L. Huang, *Normalization Techniques in Deep Learning*, Synthesis Lectures on Computer
Vision, https://doi.org/10.1007/978-3-031-14595-7_3

as constant during backpropagation. In [2], $\hat{\mu}$ is estimated by running averages. Wiesler et al. [6] also considered centering the activations to improve the performance of DNNs, reformulating the centering normalization by re-parameterization. This can be viewed as a pre-conditioning method. They also used running average to estimate $\hat{\mu}$ based on the mini-batch activations. One interesting observation in [6] is that the scaling operation does not yield improvements in this case. One likely reason is that the population statistics estimated by running average are not accurate, and thus cannot adequately exploit the advantages of standardization.

Desjardins et al. [7] proposed to whiten the activations using the population statistics as:

$$\hat{\mathbf{x}} = \widehat{\Sigma}^{-\frac{1}{2}}(\mathbf{x} - \hat{\mu}), \tag{3.2}$$

where $\widehat{\Sigma}^{-\frac{1}{2}}$ is the population statistics of the whitening matrix. One difficulty is to accurately estimate $\widehat{\Sigma}^{-\frac{1}{2}}$. In [7, 8], $\widehat{\Sigma}^{-\frac{1}{2}}$ is updated over T intervals, and the whitening matrix is further pre-conditioned by one hyperparameter ϵ, which balances the natural gradient (produced by the whitened activations) and the naive gradient. With these two techniques, the networks with whitened activations can be trained by finely adjusting T/ϵ. Luo [8] investigated the effectiveness of whitening the activations (pre-whitening) and pre-activations (post-whitening). They also addressed the computational issues by using online singular value decomposition (SVD) when calculating the whitening matrix.

Although several improvements in performance have been achieved, normalization by population statistics still faces some drawbacks. The main disadvantage is the training instability, which can be caused by the inaccurate estimation of population statistics: (1) These methods usually use a limited number of data samples to estimate the population statistics, due to computational concerns. (2) Even if full data is available and an accurate estimation is obtained for a current iteration, the activation distribution (and thus the population statistics) will change due to the updates of the weight matrix, which is known as internal covariant shift (ICS) [1]. (3) Finally, an inaccurate estimation of population statistics will be amplified as the number of layers increases, so these methods are not suitable for large-scale networks. As pointed out in [7, 8], additional batch normalization [1] is needed to stabilize the training for large-scale networks.

3.2   Local Statistics in a Sample

There is another line of work for normalizing the internal representation in computer vision communities. These methods boost contrast in the regions where it is low or moderate, while leaving it unchanged where it is high. One important characteristic is that the statistics for each example/position are calculated over the neighboring regions and thus vary. This kind of normalization is called *local normalization* [9]. Given a feature map $\mathbf{X} \in \mathbb{R}^{d \times h \times w}$, Jarrett et al. [10] propose local contrast normalization (LCN) to standardize each example's feature

X_{cij}—where c refers to channel c and i, j refer to spatial position of the feature—by the statistics calculated by its neighbors in a window of size 9×9, as:

$$\widehat{X}_{cij} = \frac{X_{cij} - \sum_{pq} w_{pq} X_{c,i+p,j+q}}{max(\delta, \sigma_{ij})} \tag{3.3}$$

where w_{pq} is a Gaussian weighting window (9×9) normalized so that $\sum_{pq} w_{pq} = 1$ and σ_{ij} is the weighted standard deviation of all features over a small spatial neighborhood (9×9). For each sample, the constant δ is set to the mean of σ_{ij}. Local response normalization (LRN) [11] proposed to scale the activation as:

$$\widehat{X}_{cij} = \frac{X_{cij}}{(k + \alpha \sum_{s=max(0,c-n/2)}^{min(d-1,c+n/2)} X_{sij}^2)^{\beta}} \tag{3.4}$$

where the sum runs over n 'adjacent' kernel maps at the same spatial position (h, w), and the constants k, n, α and β are hyper-parameters. Local context normalization [9] extends the neighborhood partition, where the normalization is performed within a window of size $p \times q$, for groups of filters with a size predefined by the number of channels per group (c_groups) along the channel axis. Divisive normalization (DN) [12] generalizes the neighborhood partition of these local normalization methods as the choices of the summation and suppression fields [12, 13].

There are two main characteristics of the local normalization methods. The first one is that the normalization operation is performed only on the feature maps of one example. The second is that the statistics for each position of feature maps are calculated over its neighboring regions and are varying.

The local normalization methods have several good advantages for training. For example, the training is more stable due to back-propagating through the normalization transformation. Besides, the model is capable to learn the visual contrast invariant property, which may benefit generalization. However, these methods have several limits. The motivation of the methods are specific to visual data or their feature maps. These methods also may change the representation ability and remove the information for discriminative classification. Besides, it is not clear whether they benefit optimization, because they do not address normalization over population statistics, which is theoretically motivated, as discussed in Chap. 2.

3.3 Batch Normalization

Batch normalization (BN) [1] paved the way to viewing normalization statistics as functions over mini-batch inputs, and addressing backpropagation through normalization operations. Let x denote the activation for a given neuron in one layer of a DNN. BN standardizes the neuron within m mini-batch data $\mathcal{B} = \{x^{(i)}\}_{i=1}^m$ by:

$$\hat{x}^{(i)} = \frac{x^{(i)} - \mu_B}{\sqrt{\sigma_B^2 + \epsilon}}, \tag{3.5}$$

where $\epsilon > 0$ is a small number to prevent numerical instability, and $\mu_B = \frac{1}{m}\sum_{i=1}^{m} x^{(i)}$ and $\sigma_B^2 = \frac{1}{m}\sum_{i=1}^{m}(x^{(i)} - \mu_B)^2$ are the mean and variance, respectively.[1] During inference, the population statistics $\{\hat{\mu}, \hat{\sigma}^2\}$ are required for deterministic inference, and they are usually calculated by running average over the training iterations, as follows:

$$\begin{cases} \hat{\mu} = (1 - \alpha)\hat{\mu} + \alpha\mu_B, \\ \hat{\sigma}^2 = (1 - \alpha)\hat{\sigma}^2 + \alpha\sigma_B^2, \end{cases} \tag{3.6}$$

where α is a scalar ranging in $[0, 1]$.

Compared to the normalization methods based on population statistics, introduced in Sect. 3.1, this normalization strategy provides several advantages: (1) It avoids using the population statistics to normalize the activations, thus avoiding the instability caused by inaccurate estimations. (2) The normalized output for each mini-batch has a zero-mean and unit-variance constraint that stabilizes the distribution of the activations, and thus benefits training. For more discussions please refer to the subsequent Chap. 9. Here, we address that the mini-batch mean μ_B and variance σ_B^2 of BN are functions of layer input $x^{(i)}$, and it is essential to back-propagate though them. To be specific, given the gradient w.r.t. the normalized output $\frac{\partial \mathcal{L}}{\partial \hat{x}^{(i)}}$ during back-propagation, the gradient w.r.t. the input $\frac{\partial \mathcal{L}}{\partial x^{(i)}}$ is calculated as follows:

$$\frac{\partial \mathcal{L}}{\partial \sigma_B^2} = \sum_{i=1}^{m} \frac{\partial \mathcal{L}}{\partial \hat{x}^{(i)}} \cdot (x^{(i)} - \mu_B) \cdot \frac{-1}{2}(\sigma_B^2 + \epsilon)^{-3/2}$$

$$\frac{\partial \mathcal{L}}{\partial \mu_B} = \sum_{i=1}^{m} \frac{\partial \mathcal{L}}{\partial \hat{x}^{(i)}} \cdot \frac{-1}{\sqrt{\sigma_B^2 + \epsilon}} + \frac{\partial \mathcal{L}}{\partial \sigma_B^2} \cdot \frac{\sum_{i=1}^{m} -2(x^{(i)} - \mu_B)}{m}$$

$$\frac{\partial \mathcal{L}}{\partial x^{(i)}} = \frac{\partial \mathcal{L}}{\partial \hat{x}^{(i)}} \cdot \frac{1}{\sqrt{\sigma_B^2 + \epsilon}} + \frac{\partial \mathcal{L}}{\partial \sigma_B^2} \cdot \frac{2(x^{(i)} - \mu_B)}{m} + \frac{\partial \mathcal{L}}{\partial \mu_B} \cdot \frac{1}{m} \tag{3.7}$$

Due to the constraints introduced by standardization, BN also uses an additional learnable scale parameter $\gamma \in \mathbb{R}$ and shift parameter $\beta \in \mathbb{R}$ to recover a possible reduced representation capacity [1]:

$$\tilde{x} = \gamma\hat{x} + \beta. \tag{3.8}$$

In this book, we also refer to the scale parameter and shift parameter as affine parameters.

The formulation of BN in Eq. 3.5 is based on the multiple layer perceptron (MLP), for better showing its motivation in optimization. Given a convolutional input for visual data, BN additionally proposes the normalization to obey the convolutional property—so

---

[1] Note that here μ_B and σ_B^2 are functions over the mini-batch data B.

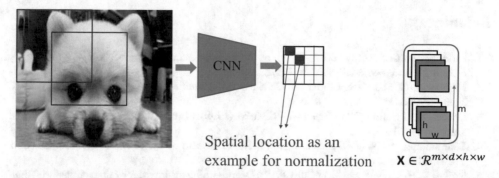

Spatial location as an
example for normalization $\mathbf{X} \in \mathcal{R}^{m \times d \times h \times w}$

Fig. 3.1 Illustration of the BN for convolutional input. We can find that the activation in the feature map is essentially the response of the corresponding region in the original image

that different elements of the same feature map, at different locations, are normalized in the same way. As illustrated in Fig. 3.1, the activation in the feature map is essentially the response of the corresponding region in the original image. Therefore, the spatial location can be viewed as an example, following the convolution property. In summary, given a mini-batch convolutional input $\mathbf{X} \in \mathbb{R}^{d \times m \times h \times w}$, where d, h and w are the channel number, height and width of the feature maps, respectively. BN jointly normalizes all the activations in a mini-batch, over all locations.

BN addresses to normalize the activation over mini-batch data for better conditioning, following the motivation from the optimization perspective. More importantly, BN also addresses to back-propagate through the transformation, like the local normalization methods, which can stabilize the training. BN can be wrapped as an independent module and plugged in the neural networks, which is very convenient to use. BN has been shown to be a milestone in the deep learning community [14–16]. It is widely used in different networks [14, 15, 17–22] and various applications [23]. However, despite its great success in deep learning, BN still faces several issues in particular contexts: (1) The inconsistent operation of BN between training and inference limits its usage in complex networks (e.g., recurrent neural networks (RNNs) [24–26]) and tasks [27–29]; (2) BN suffers from the small-batch-size problem—its error increases rapidly as the batch size becomes smaller [16]. To address BN's weaknesses and further extend its functionality, plenty of works related to feature normalization have been proposed.

In the following chapters for illustrate normalizing activation methods, we first propose a framework to describe normalizing-activations-as-function methods and review the basic single-mode normalization methods, which ensures that the normalized output has a single-mode (Gaussian) distribution. We then introduce the approaches that extend single-mode method to multiple modes, and that further combine different normalization methods. Lastly, we discuss the more robust estimation methods that address the small-batch-size problem of BN.

References

1. Ioffe, S. and C. Szegedy (2015). Batch normalization: Accelerating deep network training by reducing internal covariate shift. In *ICML*.
2. Montavon, G. and K.-R. Müller (2012). *Deep Boltzmann Machines and the Centering Trick*, Volume 7700.
3. LeCun, Y., L. Bottou, G. B. Orr, and K.-R. Muller (1998). Efficient backprop. In *Neural Networks: Tricks of the Trade*.
4. Schraudolph, N. N. (1998). Accelerated gradient descent by factor-centering decomposition. Technical report.
5. Raiko, T., H. Valpola, and Y. LeCun (2012). Deep learning made easier by linear transformations in perceptrons. In *AISTATS*.
6. Wiesler, S., A. Richard, R. Schlüter, and H. Ney (2014). Mean-normalized stochastic gradient for large-scale deep learning. In *ICASSP*.
7. Desjardins, G., K. Simonyan, R. Pascanu, and k. kavukcuoglu (2015). Natural neural networks. In *NeurIPS*.
8. Luo, P. (2017). Learning deep architectures via generalized whitened neural networks. In *ICML*.
9. Ortiz, A., C. Robinson, M. Hassan, D. Morris, O. Fuentes, C. Kiekintveld, and N. Jojic (2019). Local context normalization: Revisiting local normalization. *arXiv preprint* arXiv:1912.05845.
10. Jarrett, K., K. Kavukcuoglu, M. Ranzato, and Y. LeCun (2009). What is the best multi-stage architecture for object recognition? In *ICCV*.
11. Krizhevsky, A., I. Sutskever, and G. E. Hinton (2012). Imagenet classification with deep convolutional neural networks. In *NeurIPS*, pp. 1097–1105.
12. Ren, M., R. Liao, R. Urtasun, F. H. Sinz, and R. S. Zemel (2017). Normalizing the normalizers: Comparing and extending network normalization schemes. In *ICLR*.
13. Siwei Lyu and E. P. Simoncelli (2008). Nonlinear image representation using divisive normalization. In *CVPR*.
14. He, K., X. Zhang, S. Ren, and J. Sun (2016a). Deep residual learning for image recognition. In *CVPR*.
15. Szegedy, C., V. Vanhoucke, S. Ioffe, J. Shlens, and Z. Wojna (2016). Rethinking the inception architecture for computer vision. In *CVPR*.
16. Wu, Y. and K. He (2018). Group normalization. In *ECCV*.
17. Szegedy, C., W. Liu, Y. Jia, P. Sermanet, S. Reed, D. Anguelov, D. Erhan, V. Vanhoucke, and A. Rabinovich (2015). Going deeper with convolutions. In *CVPR*.
18. Simonyan, K. and A. Zisserman (2015). Very deep convolutional networks for large-scale image recognition. In *ICLR*.
19. Zagoruyko, S. and N. Komodakis (2016). Wide residual networks. In *BMVC*.
20. He, K., X. Zhang, S. Ren, and J. Sun (2016b). Identity mappings in deep residual networks. In *ECCV*.
21. Huang, G., Z. Liu, and K. Q. Weinberger (2017). Densely connected convolutional networks. In *CVPR*.
22. Xie, S., R. B. Girshick, P. Dollár, Z. Tu, and K. He (2017). Aggregated residual transformations for deep neural networks. In *CVPR*.
23. Bronskill, J., J. Gordon, J. Requeima, S. Nowozin, and R. E. Turner (2020a). Tasknorm: Rethinking batch normalization for meta-learning. In *ICML*.
24. Cooijmans, T., N. Ballas, C. Laurent, and A. C. Courville (2017). Recurrent batch normalization. In *ICLR*.
25. Laurent, C., G. Pereyra, P. Brakel, Y. Zhang, and Y. Bengio (2016). Batch normalized recurrent neural networks. In *ICASSP*.

26. Ba, L. J., R. Kiros, and G. E. Hinton (2016). Layer normalization. *arXiv preprint* arXiv:1607.06450.
27. Salimans, T., I. Goodfellow, W. Zaremba, V. Cheung, A. Radford, X. Chen, and X. Chen (2016). Improved techniques for training GANs. In *NeurIPS*.
28. Kurach, K., M. Lučić, X. Zhai, M. Michalski, and S. Gelly (2019). A large-scale study on regularization and normalization in GANs. In *ICML*.
29. Bhatt, A., M. Argus, A. Amiranashvili, and T. Brox (2019). Crossnorm: Normalization for off-policy TD reinforcement learning. *arXiv preprint* arXiv:1902.05605.

A Framework for Normalizing Activations as Functions

We propose a framework to describe normalizing-activations-as-function methods in Algorithm 4.1. We divide the normalizing-activations-as-function framework into three abstract processes: normalization area partitioning (NAP), normalization operation (NOP), and normalization representation recovery (NRR). We consider the more general mini-batch (of size m) activations in a convolutional layer $\mathsf{X} \in \mathbb{R}^{d \times m \times h \times w}$, where d, h and w are the channel number, height and width of the feature maps, respectively.[1] NAP transforms the activations X into $X \in \mathbb{R}^{S_1 \times S_2}$[2], where S_2 indexes the set of samples used to compute the estimators. NOP denotes the specific normalization operation (see main operations in Sect. 1.1) on the transformed data X. NRR is used to recover the possible reduced representation capacity.

Algorithm 4.1 Framework of normalizing activations as functions.

1: **Input**: mini-batch inputs $\mathsf{X} \in \mathbb{R}^{d \times m \times h \times w}$.
2: **Output**: $\widetilde{\mathsf{X}} \in \mathbb{R}^{d \times m \times h \times w}$.
3: Normalization area partitioning: $X = \Pi(\mathsf{X})$.
4: Normalization operation: $\widehat{X} = \Phi(X)$.
5: Normalization representation recovery: $\widetilde{X} = \Psi(\widehat{X})$.
6: Reshape back: $\widetilde{\mathsf{X}} = \Pi^{-1}(\widetilde{X})$.

Take BN as an example. BN considers each spatial position in a feature map as a sample [3, 4] and the NAP is:

$$X = \Pi_{BN}(\mathsf{X}) \in \mathbb{R}^{d \times mhw}, \tag{4.1}$$

---

[1] Note that the convolutional activation is reduced to the MLP activation, when setting $h = w = 1$.
[2] NAP can be implemented by the reshape operation of PyTorch [1] or Tensorflow [2].

© The Author(s), under exclusive license to Springer Nature Switzerland AG 2022
L. Huang, *Normalization Techniques in Deep Learning*, Synthesis Lectures on Computer Vision, https://doi.org/10.1007/978-3-031-14595-7_4

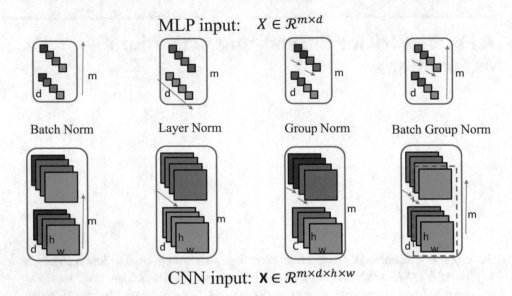

MLP input: $X \in \mathcal{R}^{m \times d}$

Batch Norm   Layer Norm   Group Norm   Batch Group Norm

CNN input: $\mathbf{X} \in \mathcal{R}^{m \times d \times h \times w}$

Fig. 4.1 Main variants of normalization by using different NAP, which apply to inputs of MLP and CNN

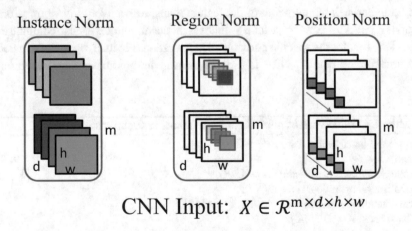

Instance Norm   Region Norm   Position Norm

CNN Input: $X \in \mathcal{R}^{m \times d \times h \times w}$

Fig. 4.2 Main variants of normalization by using different NAP, which are specific to CNN input

which means that the statistics are calculated along the batch, height, and width dimensions. The NOP is the standardization, represented in the form of a matrix as:

$$\widehat{X} = \Phi_{SD}(X) = \Lambda^{-\frac{1}{2}}(X - \mu \mathbf{1}^T). \tag{4.2}$$

Here, $\mu = \frac{1}{mhw} X \mathbf{1}$ is the mean of data samples, $\mathbf{1}$ is a column vector of all ones, and $\Lambda_d = \mathrm{diag}(\sigma_1^2, \ldots, \sigma_d^2) + \epsilon I$, where σ_j^2 is the variance over data samples for the j-th neuron/channel. The NRR is the affine transformation with channel-wise learnable affine parameters $\gamma, \beta \in \mathbb{R}^d$, defined as:

$$\widetilde{X} = \Psi_{AF}(\widehat{X}) = \widehat{X} \odot (\gamma \mathbf{1}^T) + (\beta \mathbf{1}^T), \tag{4.3}$$

where $\odot$ indicates the Hadamard product.

In the following sections, we will discuss the research progress along these three lines.

4.1   Normalization Area Partitioning

In this section, we will introduce normalization methods with different NAPs (see Figs. 4.1 and 4.2 for illustrative examples). Here, the default NOP is the standardization operation (Eq. 4.2), and the NRR is the affine transform (Eq. 4.3), unless otherwise stated.

Layer Normalization (LN) [5] proposes to standardize the layer input within the neurons for each training sample, to avoid the drawbacks of normalization along batch dimensions. Specifically, the NAP of LN is $X = \Pi_{LN}(X) \in \mathbb{R}^{m \times dhw}$, where the normalization statistics are calculated along the channel, height and width dimensions. LN has the same formulation during training and inference, and is extensively used in NLP tasks [6–8].

Group normalization (GN) [9] generalizes LN, dividing the neurons into groups and standardizing the layer input within the neurons of each group for each sample independently. Specifically, the NAP of GN is $X = \Pi_{GN}(X) \in \mathbb{R}^{mg \times shw}$, where g is the group number and $d = gs$. LN is clearly a special case of GN with $g = 1$. By changing the group number g, GN is more flexible than LN, enabling it to achieve good performance on visual tasks limited to small-batch-size training (e.g., object detection and segmentation [9]).

Batch group normalization (BGN) [10] expands the grouping mechanism of GN from being over only channels to being over both channels and batch dimensions. The NAP of BGN is $X = \Pi_{BGN}(X) \in \mathbb{R}^{g_m g \times s_m shw}$, where $m = g_m s_m$. BGN also normalizes over batch dimensions and needs to estimate the population statistics, similar to BN in Eq. 3.6. However, the group mechanism adds 'examples' for normalization, thus relieving the small-batch problem of BN to some degree.

The methods described above all apply to not only convolutional input for visual data, but also the fully-connected input where $h = w = 1$. There are also normalization methods designed for visual data, being specific to convolutional input (Fig. 4.2). Instance normalization (IN) [11] proposes to normalize each single image to remove instance-specific contrast information. Specifically, the NAP of IN is $X = \Pi_{IN}(X) \in \mathbb{R}^{md \times hw}$. Due to its ability to remove style information from the inputs, IN is widely used in image style transfer tasks [12–14]. Region normalization (RN) [15] is an extension from instance normalization, where it divides spatial pixels into different regions according to the input mask, and standardizes the activation in each region. Note that the NRR shown in RN is a set of learnable affine parameters and each affine parameters is separately for each region. Region normalization is designed for the image inpainting tasks.

Position normalization (PN) [16] standardizes the activations at each position independently across the channels. The NAP of PN is $X = \Pi_{PN}(X) \in \mathbb{R}^{mhw \times d}$. PN is designed to deal with spatial information, and has the potential to enhance the performance of generative models [16, 17].

4.2 Normalization Operation

As previously discussed, current normalization methods usually use a standardization operation. However, other operations can also be used to normalize the data. In this part, we divide these operations into three categories: (1) Extending standardization towards the whitening operation, which is a more general operation; (2) Variations of standardization; (3) Reduced standardizations that use only centering or scaling for some special situations. Unless otherwise stated, the NAP is Π_{BN}, the data transferred after the NAP is denoted as $X \in \mathbb{R}^{d \times m}$, and the NRR is the affine transform as shown in Eq. 4.3.

4.2.1 Beyond Standardization Towards Whitening

As discussed in Chap. 2, the ideal normalization operation for improving the conditioning of optimization is whitening. [18] proposed decorrelated BN, which extends BN to batch whitening (BW). The NOP of BW is whitening, represented as:

$$\widehat{X} = \Phi_W(X) = \Sigma^{-\frac{1}{2}}(X - \mu \mathbf{1}^T). \tag{4.4}$$

Here, $\Sigma^{-\frac{1}{2}}$ is the whitening matrix, which is calculated from the corresponding mini-batch covariance matrix $\Sigma = \frac{1}{m}(X - \mu \mathbf{1}^T)(X - \mu \mathbf{1}^T)^T + \epsilon I$. Note that the mean and whitening matrix are also functions over mini-batch input and we need to back-propagate through them.

One main challenge for extending standardization to whitening is how to back-propagate through the inverse square root of a matrix. This can be achieved by using matrix differential calculus [19], as proposed in [18]. There are an infinite number of possible whitening matrices, as shown in [20], since any whitened data with a rotation is still whitened. Here, we provide a general view of batch whitening methods during training shown in Algorithm 4.2, and their corresponding back-propagation in Algorithm 4.3. One interesting question is the choice of how to compute the whitening matrix $\Sigma^{-\frac{1}{2}}$. Here, we mainly introduce the whitening transformations based on PCA, Zero-phase component analysis (ZCA) and Cholesky decomposition (CD), since these three transformations have shown significant differences in performance when used in training DNNs [18, 21], even though they can equivalently improve the conditioning.

PCA Whitening uses $\Sigma_{PCA}^{-\frac{1}{2}} = \widetilde{\Lambda}^{-\frac{1}{2}} D^T$, where $\widetilde{\Lambda} = \text{diag}(\lambda_1, \ldots, \lambda_d)$ and $D = [\mathbf{d}_1, \ldots, \mathbf{d}_d]$ are the eigenvalues and associated eigenvectors of Σ, i.e. $\Sigma = D \widetilde{\Lambda} D^T$.

Algorithm 4.2 A general view of batch whitening algorithms.

1: **Input**: mini-batch inputs $X \in \mathbb{R}^{d \times m}$.
2: **Output**: $\widehat{X} \in \mathbb{R}^{d \times m}$.
3: calculate mini-batch mean: $\mu = \frac{1}{m} X \cdot \mathbf{1}$.
4: calculate centered activation: $X_C = X - \mu \cdot \mathbf{1}^T$.
5: Calculate covariance matrix: $\Sigma = \frac{1}{m} X_C X_C^T + \epsilon I$.
6: Calculate whitening matrix: $\Sigma^{-\frac{1}{2}} = \phi_1(\Sigma)$.
7: Calculate whitened output: $\widehat{X} = \Sigma^{-\frac{1}{2}} X_C$.

Algorithm 4.3 Back-propagation of batch whitening algorithms.

1: **Input**: mini-batch gradients $\frac{\partial \mathcal{L}}{\partial \widehat{X}} \in \mathbb{R}^{d \times m}$.

2: **Output**: $\frac{\partial \mathcal{L}}{\partial X} \in \mathbb{R}^{d \times m}$.

3: Calculate: $\frac{\partial \mathcal{L}}{\partial \Sigma^{-\frac{1}{2}}} = \frac{\partial \mathcal{L}}{\partial \widehat{X}} X_C^T$.

4: Calculate: $\frac{\partial \mathcal{L}}{\partial \Sigma} = \frac{\partial \mathcal{L}}{\partial \Sigma^{-\frac{1}{2}}} \frac{\partial \phi_1(\Sigma)}{\partial \Sigma}$.

5: Calculate: $\frac{\partial \mathcal{L}}{\partial X_C} = (\Sigma^{-\frac{1}{2}})^T \frac{\partial \mathcal{L}}{\partial \widehat{X}} + \frac{1}{m}(\frac{\partial \mathcal{L}}{\partial \Sigma} + \frac{\partial \mathcal{L}}{\partial \Sigma}^T) X_C$.

6: Calculate: $\frac{\partial \mathcal{L}}{\partial X} = \frac{\partial \mathcal{L}}{\partial X_C}(I - \frac{1}{m} \mathbf{1}\mathbf{1}^T)^T$.

Under this transformation, the variables are first rotated by the eigen-matrix (D) of the covariance, then scaled by the square root inverse of the eigenvalues ($\widetilde{\Lambda}^{-\frac{1}{2}}$). PCA whitening over batch data suffers significant instability in training DNNs, and hardly converges, due to the so called stochastic axis swapping (SAS), as explained in [18], which we will illustrate in Sect. 9.3. The backward pass of PCA whitening is:

$$\frac{\partial \mathcal{L}}{\partial \widetilde{\Lambda}} = \left(\frac{\partial \mathcal{L}}{\partial \Sigma^{-\frac{1}{2}}_{PCA}} \right) D \left(-\frac{1}{2} \widetilde{\Lambda}^{-3/2} \right) \tag{4.5}$$

$$\frac{\partial \mathcal{L}}{\partial D} = \left(\frac{\partial \mathcal{L}}{\partial \Sigma^{-\frac{1}{2}}_{PCA}} \right)^T \widetilde{\Lambda}^{-1/2} \tag{4.6}$$

$$\frac{\partial \mathcal{L}}{\partial \Sigma} = D \left\{ \left(\mathbf{K}^T \odot \left(D^T \frac{\partial \mathcal{L}}{\partial D} \right) \right) + \left(\frac{\partial \mathcal{L}}{\partial \widetilde{\Lambda}} \right)_{diag} \right\} D^T, \tag{4.7}$$

where $\mathbf{K} \in \mathbb{R}^{d \times d}$ is 0-diagonal with $\mathbf{K}_{ij} = \frac{1}{\lambda_i - \lambda_j}[i \neq j]$, the $\odot$ operator is an element-wise matrix multiplication, and $(\frac{\partial \mathcal{L}}{\partial \Lambda})_{diag}$ sets the off-diagonal elements of $\frac{\partial \mathcal{L}}{\partial \Lambda}$ to zero. Note that

Eq. 4.5 is the formulation of back-propagation through eigenvalue decomposition, and we provide the details of derivation in Appendix A.1.

ZCA Whitening uses $\Sigma_{ZCA}^{-\frac{1}{2}} = D\widetilde{\Lambda}^{-\frac{1}{2}}D^T$, where the PCA whitened input is rotated back by the corresponding rotation matrix D. ZCA whitening works by stretching/squeezing the dimensions along the eigenvectors. It has been shown that ZCA whitening avoids the SAS and achieves better performance over standardization (used in BN) on discriminative classification tasks [18]. The backward pass of ZCA whitening is:

$$\frac{\partial \mathcal{L}}{\partial \widetilde{\Lambda}} = D^T \left(\frac{\partial \mathcal{L}}{\partial \Sigma_{ZCA}^{-\frac{1}{2}}} \right) D \left(-\frac{1}{2}\widetilde{\Lambda}^{-3/2} \right) \tag{4.8}$$

$$\frac{\partial \mathcal{L}}{\partial D} = \left(\frac{\partial \mathcal{L}}{\partial \Sigma_{ZCA}^{-\frac{1}{2}}} + \left(\frac{\partial \mathcal{L}}{\partial \Sigma_{ZCA}^{-\frac{1}{2}}} \right)^T \right) D\widetilde{\Lambda}^{-1/2} \tag{4.9}$$

$$\frac{\partial \mathcal{L}}{\partial \Sigma} = D \left\{ \left(\mathbf{K}^T \odot \left(D^T \frac{\partial \mathcal{L}}{\partial D} \right) \right) + \left(\frac{\partial \mathcal{L}}{\partial \widetilde{\Lambda}} \right)_{diag} \right\} D^T. \tag{4.10}$$

CD Whitening uses $\Sigma_{CD}^{-\frac{1}{2}} = L^{-1}$ where L is a lower triangular matrix from the Cholesky decomposition, with $LL^T = \Sigma$ [21]. This kind of whitening works by recursively decorrelating the current dimension over the previous decorrelated ones, resulting in a triangular form of its whitening matrix. CD whitening has been shown to achieve the state-of-the-art performance in training GANs, while ZCA whitening has degenerated performance. The backward pass of CD whitening is:

$$\frac{\partial \mathcal{L}}{\partial \mathbf{L}} = - \left(\Sigma_{CD}^{-\frac{1}{2}} \right)^T \frac{\partial \mathcal{L}}{\partial \Sigma_{CD}^{-\frac{1}{2}}} \left(\Sigma_{CD}^{-\frac{1}{2}} \right)^T \tag{4.11}$$

$$\frac{\partial \mathcal{L}}{\partial \Sigma} = \frac{1}{2}\mathbf{L}^{-T} \left(P \odot \mathbf{L}^T \frac{\partial \mathcal{L}}{\partial \mathbf{L}} + \left(P \odot \mathbf{L}^T \frac{\partial \mathcal{L}}{\partial \mathbf{L}} \right)^T \right) \mathbf{L}^{-1}. \tag{4.12}$$

Control the Extent of Batch Whitening

Even though batch whitening has better conditioning over BN (using standardization), there also raise several problems in practice: (1) The eigen-decomposition/singular value decomposition/Cholesky decomposition used to calculate the whitening matrix introduce significant computational cost, especially running on GPU; (2) The fully whitening operation provides overmuch constraints on the whitened output, which may constrain the representation ability of the model, and we will further illustrate in Sect. 9.4; (3) The stochasticity introduced by normalization along batch dimension will be significantly large, and the estimation of population statistics will be difficult, if an undersized data batch is provided for

---

Algorithm 4.4 Whitening activations using Newton's iteration.

1: **Input**: Σ.

2: **Output**: $\Sigma_{ItN}^{-\frac{1}{2}}$.

3: $\Sigma_N = \frac{\Sigma}{tr(\Sigma)}$.

4: $P_0 = I$.

5: **for** $k = 1$ to T **do**

6:    $P_k = \frac{1}{2}(3P_{k-1} - P_{k-1}^3 \Sigma_N)$

7: **end for**

8: $\Sigma_{ItN}^{-\frac{1}{2}} = P_T / \sqrt{tr(\Sigma)}$

---

training, and we will further illustrate in Sect. 9.3. Therefore, it is essential to control the extent of the whitening in practice, which provides a solution for trade-off between the advantages and dis-advantages of whitening operation.

To relieve these problems, the group based BW—where features are divided into groups and whitening is performed within each one—was proposed [18, 22] to control the extent of the whitening. One interesting property is that group-based BW reduces to BN if the channel number in each group is set to 1. Besides, group-based BW also has the added benefit of reducing the computational cost of whitening.

Reference [23] proposed iterative normalization (IterNorm) to improve the computational efficiency and numerical stability of ZCA whitening, since it can avoid eigen-decomposition or SVD by employing Newton's iteration for approximating the whitening matrix $\Sigma^{-\frac{1}{2}}$. IterNorm is shown in Algorithm 4.4, where T is the iteration number and $tr(\cdot)$ denotes the trace of a matrix. Given $\frac{\partial \mathcal{L}}{\partial \Sigma_{ItN}^{-\frac{1}{2}}}$, the back-propagation is:

$$\frac{\partial \mathcal{L}}{\partial P_T} = \frac{1}{\sqrt{tr(\Sigma)}} \frac{\partial \mathcal{L}}{\partial \Sigma_{ItN}^{-\frac{1}{2}}}$$

$$\frac{\partial \mathcal{L}}{\partial \Sigma_N} = -\frac{1}{2} \sum_{k=1}^{T} \left(P_{k-1}^3\right)^T \frac{\partial \mathcal{L}}{\partial P_k}$$

$$\frac{\partial \mathcal{L}}{\partial \Sigma} = \frac{1}{tr(\Sigma)} \frac{\partial \mathcal{L}}{\partial \Sigma_N} - \frac{1}{(tr(\Sigma))^2} tr\left(\frac{\partial \mathcal{L}}{\partial \Sigma_N}^T \Sigma\right) I$$

$$- \frac{1}{2(tr(\Sigma))^{3/2}} tr\left(\left(\frac{\partial \mathcal{L}}{\partial \Sigma_{ItN}^{-\frac{1}{2}}}\right)^T P_T\right) I. \tag{4.13}$$

Here, $\frac{\partial \mathcal{L}}{\partial P_k}$ can be calculated by the following iterations:

$$\frac{\partial \mathcal{L}}{\partial P_{k-1}} = \frac{3}{2} \frac{\partial L}{\partial P_k} - \frac{1}{2} \frac{\partial \mathcal{L}}{\partial P_k} \left(P_{k-1}^2 \Sigma_N \right)^T - \frac{1}{2} \left(P_{k-1}^2 \right)^T \frac{\partial \mathcal{L}}{\partial P_k} \Sigma_N^T$$

$$- \frac{1}{2} \left(P_{k-1} \right)^T \frac{\partial \mathcal{L}}{\partial P_k} \left(P_{k-1} \Sigma_N \right)^T, \quad k = T, \dots, 1. \tag{4.14}$$

A similar idea was also used in [22] by coupled Newton-Schulz iterations [24] for whitening. One interesting property of IterNorm is that it stretches the dimensions along the eigenvectors progressively, so that the associated eigenvalues converge to 1 after normalization. Therefore, IterNorm can effectively control the extent of whitening by its iteration number.

There also exist works that impose extra penalties on the loss function to obtain approximately whitened activations [25–29], or exploit the whitening operation to improve the network's generalization [30, 31]. Please refer to these papers for details.

Here, we discuss the advantages and disadvantages of BW, compared to BN. One advantage of BW is that it has better conditioning over BN theoretically. Secondly, BW probably has a better generalization (the amplified stochasticity and regularization on the normalized output) by well controlling the extent. The main disadvantages of BW are the high computational cost and numerical instability when calculating the whitening matrix. These problems can be well relieved by using different methods to (approximately) calculate the whitening matrix, for example, using Newton's iteration. Another disadvantage of BW is that it amplifies the disadvantage of BN in estimating the population statistics, where the number of parameters to be estimated with BW is quadratic to the number of neurons/channels. Thus, BW requires a sufficiently large batch size to work well. This problem can be avoided by using group whitening (GW) [32], which exploits the advantages of whitening and avoids the disadvantages of normalization over batch. Specifically, given a mini-batch inputs $X \in \mathbb{R}^{d \times m \times h \times w}$, GW performs normalization as $\phi_W(\Pi_{GN}(X))$. Group whitening is different to group-based BW, in which the whitening operation is also applied within mini-batch data and the estimation of population statistics is still required. The whitening operation of GW is performed on a single example and it has the consistent normalization operation during training and inference. Note that group-based BW is reduced to BN if the channel number in each group $c = 1$, while GW is reduced to GN if the group number $g = 1$.

4.2.2 Variations of Standardization

There are several variations of the standardization operation for normalizing the activations. As an alternative to the L^2 normalization used to control the activation scale in a BN layer [3], the L^1 normalization was proposed in [33–35] for standardization. Specifically, the dimension-wise standardization deviation of L^1 normalization is: $\sigma = \frac{1}{m} \sum_{i=1}^{m} |x^{(i)} - u|$. Note that L^2 normalization is made equivalent to L^1 normalization (under mild assumptions) by multiplying it by a scaling factor $\sqrt{\frac{\pi}{2}}$ [34, 35].

L^1 normalization can improve numerical stability in a low-precision implementation, as well as provide computational and memory benefits, over L^2 normalization. The merits of

L^1 normalization originate from the fact that it avoids the costly square and root operations of L^2 normalization. Specifically, [34] showed that the proposed sign and absolute operations in L^1 normalization can achieve a $1.5\times$ speedup and reduce the power consumption by 50% on an FPGA platform. Similar merits also exist when using L^∞, as discussed in [35], where the standardization deviation is: $\sigma = \max_i |x^{(i)}|$. The more generalized L^p was investigated in [33, 35], where the standardization deviation is: $\sigma = \frac{1}{m} \sqrt[p]{\sum_{i=1}^{m} (x^{(i)})^p}$.

Reference [36] proposed generalized batch normalization (GBN), in which the mean statistics for centering and the deviation measures for scaling are more general operations, the choice of which can be guided by the risk theory. They provided some optional asymmetric deviation measures for networks with ReLU, such as, the Right Semi-Deviation (RSD) [36].

4.2.3 Reduced Standardization

As stated in Sect. 1.1, the standardizing operation usually includes centering and scaling. However, some works only consider one or the other, for specific situations. Note that either centering or scaling along the batch dimension can benefit optimization, as shown in [3, 37]. Besides, the scaling operation is important for scale-invariant learning, and has been shown useful for adaptively adjusting the learning rate to stabilize training [38].

Reference [39] proposed mean-only batch normalization (MoBN), which only performs centering along the batch dimension, and works well when combined with weight normalization [39]. Reference [40] proposed to perform scaling only in BN for small-batch-size training, which also works well when combined with weight centralization. Reference [41] also proposed to perform scaling only in BN to improve the performance for NLP tasks.

Reference [17] proposed pixel normalization, where the scaling only operation is performed along the channel dimension for each position of each image. This works like PN [16] but only uses the scaling operation. Pixel normalization works well for GANs [16] when used in the generator.

Reference [42] hypothesized that the re-centering invariance produced by centering in LN [5] is dispensable and proposed to perform scaling only for LN, which is referred to as root mean square layer normalization (RMSLN). RMSLN takes into account the importance of the scale-invariant property for LN. RMSLN works as well as LN on NLP tasks but reduces the running time [42]. This idea was also used in the variance-only layer normalization for the click-through rate (CTR) prediction task [43]. Reference [44] also proposed to perform scaling along the channel dimension (like RMSLN) in the proposed online normalization to stabilize the training.

Singh and Krishnan proposed filter response normalization (FRN) [45], which performs the scaling-only operation along each channel (filter response) for each sample independently. This is similar to IN [11] but without performing the centering operation. The motivation is that the benefits of centering for normalization schemes that are batch independent are not really justified.

4.3   Normalization Representation Recovery

Normalization constrains the distribution of the activations, which can benefit optimization. However, these constraints may hamper the representation ability, so an additional affine transformation is usually used to recover the possible representation, as shown in Eq. 4.3. Besides, the normalization statistics (e.g., mean, covariance) have certain property for visual data. So normalization operation can remove these properties and one may add these properties back if they are required for specific tasks. In this view, the extra NRR is also advocated after normalization. There are also other options for constructing the NRR.

Reference [21] proposed a coloring transformation to recover the possible loss in representation ability caused by the whitening operation, which is formulated as:

$$\widetilde{X} = \Psi_{LR}(\widehat{X}) = \widehat{X}\mathbf{W} + (\beta\mathbf{1}^T), \tag{4.15}$$

where $\mathbf{W}$ is a $d \times d$ learnable matrix. The coloring transformation can be viewed as a linear layer in neural networks.

In Eq. 4.3, the NRR parameters are both learnable through backpropagation. Several works have attempted to generalize these parameters by using a hypernetwork to dynamically generate them, which is formulated as:

$$\widetilde{X} = \Psi_{DC}(\widehat{X}) = \widehat{X} \odot \Gamma_{\phi^\gamma} + B_{\phi^\beta}, \tag{4.16}$$

where $\Gamma_{\phi^\gamma} \in \mathbb{R}^{d \times m}$ and $B_{\phi^\beta} \in \mathbb{R}^{d \times m}$ are generated by the subnetworks $\phi_{\theta_\gamma}^\gamma(\cdot)$ and $\phi_{\theta_\beta}^\beta(\cdot)$, respectively. The affine parameters generated depend on the original inputs themselves, making them different from the affine parameters shown in Eq. 4.3, which are learnable by backpropagation.

Reference [46] proposed dynamic layer normalization (DLN) in a long short-term memory (LSTM) architecture for speech recognition, where $\phi_{\theta_\gamma}^\gamma(\cdot)$ and $\phi_{\theta_\beta}^\beta(\cdot)$ are separate utterance-level feature extractor subnetworks, which are jointly trained with the main acoustic model. The input of the subnetworks is the output of the corresponding hidden layer of the LSTM. Similar ideas are also used in adaptive instance normalization (AdaIN) [13] and adaptive layer-instance normalization (AdaLIN) [47] for unsupervised image-to-image translation, where the subnetworks are MLPs and the inputs of the subnetworks are the embedding features produced by one encoder. Reference [48] proposed instance-level meta normalization (ILMN), which utilizes an encoder-decoder subnetwork to generate affine parameters, given the instance-level mean and variance as input. Besides, ILMN also combines the learnable affine parameters shown in Eq. 4.3. Rather than using the channel-wise affine parameters shared across spatial positions, spatially adaptive denormalization (SPADE) [49] uses the spatially dependent $\beta, \gamma \in \mathbb{R}^{d \times h \times w}$, which are dynamically generated by a two-layer CNN with raw images as inputs.

The mechanism for generating affine parameter Γ_{ϕ^γ} shown in Eq. 4.16 resembles the squeeze-excitation (SE) block [50], when the input of the subnetwork $\phi_{\theta_\gamma}^\gamma(\cdot)$ is X itself, i.e.,

$\Gamma_{\phi^\gamma} = \phi_{\theta_\gamma}^\gamma(X)$. Inspired by this, [51] proposed instance enhancement batch normalization (IEBN), which combines the channel-wise affine parameters in Eq. 4.3 and the instance-specific channel-wise affine parameters in Eq. 4.16 using SE-like subnetworks, with fewer parameters. IEBN can effectively regulate noise by introducing instance-specific information for BN. This idea is further generalized by attentive normalization [52], where the affine parameters are modeled by K mixture components.

Rather than using a subnetwork to generate the affine parameters, [8] proposed adaptive normalization (AdaNorm), where the affine parameters depend on the standardized output $\widehat{X}$ of layer normalization:

$$\widetilde{X} = \widehat{X} \odot \phi(\widehat{X}). \tag{4.17}$$

Here, $\phi(\widehat{X})$ used in [8] is: $\phi(\hat{\mathbf{x}}) = C(1 - k\hat{\mathbf{x}}), C = \frac{1}{d}\sum_{i=1}^{d}\hat{\mathbf{x}}_i$ and k is a constant that satisfies certain constraints. Note that $\phi(\widehat{X})$ is treated as a changing constant (not a function) and the gradient of $\phi(\widehat{X})$ is detached in the implementation [8]. We also note that the side information can be injected into the NRR operation for conditional generative models. The typical works are conditional BN (CBN) [53] and conditional IN (CIN) [12]. We will elaborate on these work in Chap. 10 when we discuss applications of normalization.

In summary, we list the main single-mode normalization methods under our proposed framework in Table 4.1. Even though extensive normalization techniques exist, BN is still preferred in practice, particularly for computer vision tasks, due to its theoretical benefits in optimization (mini-batch statistics can be viewed as a stochastic approximation to the population statistics) and potentially better generalization (see Chap. 9 for details). However, the inconsistency of BN between training and inference is a latent trouble when applying BN to the scenarios that population statistics are not well defined or difficult to be estimated. These scenarios include *non-i.i.d.* training, corrupted input, using complicated architectures (e.g., RNN), small-batch training, and so on (see Chaps. 6 and 10 for details). Batch-free normalization (e.g., LN and GN) can avoid the estimation of population statistics and use consistent operations during training and inference, which is preferentially used in these scenarios. For example, LN is commonly adopted in RNN and Transformer, particularly for NLP tasks, and GN usually works better than BN in the scenario where a small-batch training (e.g., batch size of 2 or 4) is required, particularly for large-scale/complicated datasets.

Table 4.1 Summary of the main single-mode normalization methods, based on our proposed framework for describing normalizing-activations-as-functions methods. The order is based on the time of publication

Method	NAP	NOP	NRR	Published In
Batch normalization (BN)	$\Pi_{BN}(\mathsf{X}) \in \mathbb{R}^{d \times mhw}$	Standardizing	Learnable $\gamma, \beta \in \mathbb{R}^d$	ICML, 2015
Mean-only BN	$\Pi_{BN}(\mathsf{X}) \in \mathbb{R}^{d \times mhw}$	Centering	No	NeurIPS, 2016
Layer normalization (LN)	$\Pi_{LN}(\mathsf{X}) \in \mathbb{R}^{m \times dhw}$	Standardizing	Learnable $\gamma, \beta \in \mathbb{R}^d$	Arxiv, 2016
Instance normalization (IN)	$\Pi_{IN}(\mathsf{X}) \in \mathbb{R}^{md \times hw}$	Standardizing	Learnable $\gamma, \beta \in \mathbb{R}^d$	Arxiv, 2016
L^p-Norm BN	$\Pi_{BN}(\mathsf{X}) \in \mathbb{R}^{d \times mhw}$	Standardizing with L^p-Norm divided	Learnable $\gamma, \beta \in \mathbb{R}^d$	Arxiv, 2016
Conditional IN	$\Pi_{IN}(\mathsf{X}) \in \mathbb{R}^{md \times hw}$	Standardizing	Side information	ICLR, 2017
Dynamic LN	$\Pi_{LN}(\mathsf{X}) \in \mathbb{R}^{m \times dhw}$	Standardizing	Generated $\gamma, \beta \in \mathbb{R}^d$	INTERSPEECH, 2017
Conditional BN	$\Pi_{BN}(\mathsf{X}) \in \mathbb{R}^{d \times mhw}$	Standardizing	Side information	NeurIPS, 2017
Pixel normalization	$\Pi_{PN}(\mathsf{X}) \in \mathbb{R}^{mhw \times d}$	Scaling	No	ICLR, 2018
Decorrelated BN	$\Pi_{BN}(\mathsf{X}) \in \mathbb{R}^{d \times mhw}$	ZCA whitening	Learnable $\gamma, \beta \in \mathbb{R}^d$	CVPR, 2018
Group normalization (GN)	$\Pi_{GN}(\mathsf{X}) \in \mathbb{R}^{mg \times shw}$	Standardizing	Learnable $\gamma, \beta \in \mathbb{R}^d$	ECCV, 2018
Adaptive IN	$\Pi_{IN}(\mathsf{X}) \in \mathbb{R}^{md \times hw}$	Standardizing	Generated $\gamma, \beta \in \mathbb{R}^d$	ECCV, 2018
L^1-Norm BN	$\Pi_{BN}(\mathsf{X}) \in \mathbb{R}^{d \times mhw}$	Standardizing with L^1-Norm divided	Learnable $\gamma, \beta \in \mathbb{R}^d$	NeurIPS, 2018
Whitening and coloring BN	$\Pi_{BN}(\mathsf{X}) \in \mathbb{R}^{d \times mhw}$	CD whitening	Color transformation	ICLR, 2019
Generalized BN	$\Pi_{BN}(\mathsf{X}) \in \mathbb{R}^{d \times mhw}$	General standardizing	Learnable $\gamma, \beta \in \mathbb{R}^d$	AAAI, 2019

(continued)

Table 4.1 (continued)

Method	NAP	NOP	NRR	Published In
Iterative normalization	$\Pi_{BN}(X) \in \mathbb{R}^{d \times mhw}$	ZCA whitening by Newton's iteration	Learnable $\gamma, \beta \in \mathbb{R}^d$	CVPR, 2019
Instance-level meta normalization	$\Pi_{LN}(X)/\Pi_{IN}(X)/\Pi_{GN}(X)$	Standardizing	Learnable & generated $\gamma, \beta \in \mathbb{R}^d$	CVPR, 2019
Spatially adaptive denormalization	$\Pi_{BN}(X) \in \mathbb{R}^{d \times mhw}$	Standardizing	Generated $\gamma, \beta \in \mathbb{R}^{d \times h \times w}$	CVPR, 2019
Position normalization (PN)	$\Pi_{PN}(X) \in \mathbb{R}^{mhw \times d}$	Standardizing	Learnable $\gamma, \beta \in \mathbb{R}^d$	NeurIPS, 2019
Root mean square LN	$\Pi_{LN}(X) \in \mathbb{R}^{m \times dhw}$	Scaling	Learnable $\gamma \in \mathbb{R}^d$	NeurIPS, 2019
Online normalization	$\Pi_{LN}(X) \in \mathbb{R}^{m \times dhw}$	Scaling	No	NeurIPS, 2019
Batch group normalization	$\Pi_{BGN}(X) \in \mathbb{R}^{g_m g \times s_m shw}$	Standardizing	Learnable $\gamma, \beta \in \mathbb{R}^d$	ICLR, 2020
Instance enhancement BN	$\Pi_{BN}(X) \in \mathbb{R}^{d \times mhw}$	Standardizing	Learnable & generated $\gamma, \beta \in \mathbb{R}^d$	AAAI, 2020
PowerNorm	$\Pi_{BN}(X) \in \mathbb{R}^{d \times mhw}$	Scaling	Learnable $\gamma, \beta \in \mathbb{R}^d$	ICML, 2020
Filter response normalization	$\Pi_{IN}(X) \in \mathbb{R}^{md \times hw}$	Scaling	Learnable $\gamma, \beta \in \mathbb{R}^d$	CVPR, 2020
Attentive normalization	$\Pi_{BN}(X)/\Pi_{IN}(X)/\Pi_{LN}(X)/\Pi_{GN}(X)$	Standardizing	Generated $\gamma, \beta \in \mathbb{R}^d$	ECCV, 2020
Group whitening	$\Pi_{GN}(X)$	ZCA Whitening by Newton's iteration	Learnable $\gamma, \beta \in \mathbb{R}^d$	CVPR, 2021

References

1. Paszke, A., S. Gross, S. Chintala, G. Chanan, E. Yang, Z. DeVito, Z. Lin, A. Desmaison, L. Antiga, and A. Lerer (2017). Automatic differentiation in PyTorch. In *NeurIPS Autodiff Workshop*.
2. Abadi, M., P. Barham, J. Chen, Z. Chen, A. Davis, J. Dean, M. Devin, S. Ghemawat, G. Irving, M. Isard, et al. (2016). Tensorflow: A system for large-scale machine learning. In *12th {USENIX} Symposium on Operating Systems Design and Implementation ({OSDI} 16)*.

3. Ioffe, S. and C. Szegedy (2015). Batch normalization: Accelerating deep network training by reducing internal covariate shift. In *ICML*.
4. Gülçehre, Ç. and Y. Bengio (2016). Knowledge matters: Importance of prior information for optimization. *The Journal of Machine Learning Research 17*(1), 226–257.
5. Ba, L. J., R. Kiros, and G. E. Hinton (2016). Layer normalization. *arXiv preprint* arXiv:1607.06450.
6. Vaswani, A., N. Shazeer, N. Parmar, J. Uszkoreit, L. Jones, A. N. Gomez, L. Kaiser, and I. Polosukhin (2017). Attention is all you need. In *NeurIPS*.
7. Yu, A. W., D. Dohan, M.-T. Luong, R. Zhao, K. Chen, M. Norouzi, and Q. V. Le (2018). Qanet: Combining local convolution with global self-attention for reading comprehension. In *ICLR*.
8. Xu, J., X. Sun, Z. Zhang, G. Zhao, and J. Lin (2019). Understanding and improving layer normalization. In *NeurIPS*.
9. Wu, Y. and K. He (2018). Group normalization. In *ECCV*.
10. Summers, C. and M. J. Dinneen (2020). Four things everyone should know to improve batch normalization. In *ICLR*.
11. Ulyanov, D., A. Vedaldi, and V. S. Lempitsky (2016). Instance normalization: The missing ingredient for fast stylization. *arXiv preprint* arXiv:1607.08022.
12. Dumoulin, V., J. Shlens, and M. Kudlur (2017). A learned representation for artistic style. In *ICLR*.
13. Huang, X., M. Liu, S. J. Belongie, and J. Kautz (2018). Multimodal unsupervised image-to-image translation. In *ECCV*.
14. Huang, X. and S. Belongie (2017). Arbitrary style transfer in real-time with adaptive instance normalization. In *ICCV*.
15. Yu, T., Z. Guo, X. Jin, S. Wu, Z. Chen, W. Li, Z. Zhang, and S. Liu (2020). Region normalization for image inpainting. In *AAAI*.
16. Li, B., F. Wu, K. Q. Weinberger, and S. Belongie (2019). Positional normalization. In *NeurIPS*.
17. Karras, T., T. Aila, S. Laine, and J. Lehtinen (2018). Progressive growing of GANs for improved quality, stability, and variation. In *ICLR*.
18. Huang, L., D. Yang, B. Lang, and J. Deng (2018). Decorrelated batch normalization. In *CVPR*.
19. Ionescu, C., O. Vantzos, and C. Sminchisescu (2015). Training deep networks with structured layers by matrix backpropagation. In *ICCV*.
20. Kessy, A., A. Lewin, and K. Strimmer (2018). Optimal whitening and decorrelation. *The American Statistician 72*(4), 309–314.
21. Siarohin, A., E. Sangineto, and N. Sebe (2019). Whitening and coloring transform for GANs. In *ICLR*.
22. Ye, C., M. Evanusa, H. He, A. Mitrokhin, T. Goldstein, J. A. Yorke, C. Fermuller, and Y. Aloimonos (2020). Network deconvolution. In *ICLR*.
23. Huang, L., Y. Zhou, F. Zhu, L. Liu, and L. Shao (2019). Iterative normalization: Beyond standardization towards efficient whitening. In *CVPR*.
24. Higham, N. J. (2008). *Functions of matrices: theory and computation*. SIAM.
25. Cogswell, M., F. Ahmed, R. B. Girshick, L. Zitnick, and D. Batra (2016). Reducing overfitting in deep networks by decorrelating representations. In *ICLR*.
26. Xiong, W., B. Du, L. Zhang, R. Hu, and D. Tao (2016). Regularizing deep convolutional neural networks with a structured decorrelation constraint. In *ICDM*.
27. Littwin, E. and L. Wolf (2018). Regularizing by the variance of the activations' sample-variances. In *NeurIPS*.
28. Joo, T., D. Kang, and B. Kim (2020). Regularizing activations in neural networks via distribution matching with the wasserstein metric. In *ICLR*.
29. Zhou, W., B. Y. Lin, and X. Ren (2020). IsoBN: Fine-tuning BERT with isotropic batch normalization. *arXiv preprint* arXiv:2005.02178.

30. Shao, W., S. Tang, X. Pan, P. Tan, X. Wang, and P. Luo (2020). Channel equilibrium networks for learning deep representation. In *ICML*.
31. Chen, Z., Y. Bei, and C. Rudin (2020). Concept whitening for interpretable image recognition. *arXiv preprint* arXiv:2002.01650.
32. Huang, L., Y. Zhou, L. Liu, F. Zhu, and L. Shao (2021). Group whitening: Balancing learning efficiency and representational capacity. In *CVPR*.
33. Liao, Q., K. Kawaguchi, and T. Poggio (2016). Streaming normalization: Towards simpler and more biologically-plausible normalizations for online and recurrent learning. *arXiv preprint* arXiv:1610.06160.
34. Wu, S., G. Li, L. Deng, L. Liu, Y. Xie, and L. Shi (2018). L1-norm batch normalization for efficient training of deep neural networks. *arXiv preprint* arXiv:1802.09769.
35. Hoffer, E., R. Banner, I. Golan, and D. Soudry (2018). Norm matters: efficient and accurate normalization schemes in deep networks. In *NeurIPS*.
36. Yuan, X., Z. Feng, M. Norton, and X. Li (2019). Generalized batch normalization: Towards accelerating deep neural networks. In *AAAI*.
37. LeCun, Y., L. Bottou, G. B. Orr, and K.-R. Muller (1998). Efficient backprop. In *Neural Networks: Tricks of the Trade*.
38. Arora, S., Z. Li, and K. Lyu (2019). Theoretical analysis of auto rate-tuning by batch normalization. In *ICLR*.
39. Salimans, T. and D. P. Kingma (2016). Weight normalization: A simple reparameterization to accelerate training of deep neural networks. In *NeurIPS*.
40. Yan, J., R. Wan, X. Zhang, W. Zhang, Y. Wei, and J. Sun (2020). Towards stabilizing batch statistics in backward propagation of batch normalization. In *ICLR*.
41. Shen, S., Z. Yao, A. Gholami, M. W. Mahoney, and K. Keutzer (2020). Powernorm: Rethinking batch normalization in transformers. In *ICML*.
42. Zhang, B. and R. Sennrich (2019). Root mean square layer normalization. In *NeurIPS*.
43. Wang, Z., Q. She, P. Zhang, and J. Zhang (2020). Correct normalization matters: Understanding the effect of normalization on deep neural network models for click-through rate prediction. *arXiv preprint* arXiv:2006.12753.
44. Chiley, V., I. Sharapov, A. Kosson, U. Koster, R. Reece, S. Samaniego de la Fuente, V. Subbiah, and M. James (2019). Online normalization for training neural networks. In *NeurIPS*.
45. Singh, S. and S. Krishnan (2020). Filter response normalization layer: Eliminating batch dependence in the training of deep neural networks. In *CVPR*.
46. Kim, T., I. Song, and Y. Bengio (2017). Dynamic layer normalization for adaptive neural acoustic modeling in speech recognition. In *INTERSPEECH*, pp. 2411–2415.
47. Kim, J., M. Kim, H. Kang, and K. H. Lee (2020). U-gat-it: Unsupervised generative attentional networks with adaptive layer-instance normalization for image-to-image translation. In *ICLR*.
48. Jia, S., D. Chen, and H. Chen (2019). Instance-level meta normalization. In *CVPR*.
49. Park, T., M.-Y. Liu, T.-C. Wang, and J.-Y. Zhu (2019). Semantic image synthesis with spatially-adaptive normalization. In *CVPR*.
50. Hu, J., L. Shen, and G. Sun (2018). Squeeze-and-excitation networks. In *CVPR*.
51. Liang, S., Z. Huang, M. Liang, and H. Yang (2020). Instance enhancement batch normalization: an adaptive regulator of batch noise. In *AAAI*.
52. Li, X., W. Sun, and T. Wu (2019). Attentive normalization. *ArXiv arXiv preprint* arXiv:1908.01259.
53. de Vries, H., F. Strub, J. Mary, H. Larochelle, O. Pietquin, and A. C. Courville (2017). Modulating early visual processing by language. In *NeurIPS*, pp. 6594–6604.

Multi-mode and Combinational Normalization

In previous chapters, we focused on single-mode normalization methods. In this chapter, we will introduce the methods that extend to multiple modes, as well as combinational methods.

5.1 Multiple Modes

Kalayeh and Shah [1] proposed mixture normalizing (MixNorm), which performs normalization on subregions that can be identified by disentangling the different modes of the distribution, estimated via a Gaussian mixture model (GMM). Specifically, they assume the activations $\mathbf{x} \in \mathbb{R}^d$ satisfy a GMM distribution as:

$$p(\mathbf{x}) = \sum_{k=1}^{K} \alpha_k p_k(\mathbf{x}) \ \ s.t. \forall : \alpha_k \leq 0, \sum_{k=1}^{K} \alpha_k = 1, \tag{5.1}$$

where

$$p_k(\mathbf{x}) = \frac{1}{(2\pi)^{d/2}|\Sigma_k|^{1/2}} exp\left\{ -\frac{1}{2}(\mathbf{x} - \mu_k)^T \Sigma_k^{-1}(\mathbf{x} - \mu_k) \right\} \tag{5.2}$$

represents k-th Gaussian in the mixture model $p(\mathbf{x})$. It is possible to estimate the mixture coefficient α_k and further derive the soft-assignment mechanism $\nu_k(\mathbf{x}) = \frac{\alpha_k p_k(\mathbf{x})}{\sum_{j=1}^{K} \alpha_j p_j(\mathbf{x})}$, by using the expectation-maximization (EM) [2] algorithm. Kalayeh and Shah [1] thus define the mixture normalization transform as:

$$\hat{\mathbf{x}}^{(i)} = \sum_{k=1}^{K} \frac{\nu_k(\mathbf{x}^{(i)})}{\sqrt{\alpha_k}} \cdot \frac{\mathbf{v}_k^{(i)}}{\sqrt{\mathbb{E}_{\mathbf{x} \sim \mathbb{D}}[\hat{\nu}_k(\mathbf{x}) \cdot \mathbf{v}_k^2] + \epsilon}}, \tag{5.3}$$

© The Author(s), under exclusive license to Springer Nature Switzerland AG 2022
L. Huang, *Normalization Techniques in Deep Learning*, Synthesis Lectures on Computer Vision, https://doi.org/10.1007/978-3-031-14595-7_5

where $\mathbf{v}_k^{(i)} = (\mathbf{x}^{(i)} - \mathbb{E}_{\mathbf{x}\sim\mathbb{D}}[\hat{\nu}_k(\mathbf{x}) \cdot \mathbf{x}])$ and $\hat{\nu}_k(\mathbf{x}^{(i)}) = \frac{\nu_k(\mathbf{x}^{(i)})}{\sum_{\mathbf{x}^{(j)}\in\mathbb{D}} \nu_k(\mathbf{x}^{(j)})}$ is the normalized contribution of $\mathbf{x}^{(i)}$ over data set $\mathbb{D}$ in estimating statistical measures of the k-th Gaussian component. MixNorm requires a two-stage process, where the GMM is first fitted by expectation-maximization (EM) [2] with K-means++ [3] for initialization, and the normalization is then performed on samples with respect to the estimated parameters. MixNorm is not fully differentiable due to the K-means++ and EM iterations.

Deecke et al. [4] proposed mode normalization (ModeNorm), which also extends the normalization to more than one mean and variance to address the heterogeneous nature of complex datasets. MN is formulated in a mixture of experts (MoE) framework, where a set of simple gate functions is introduced to assign one example to groups with a given probability. In particular, they introduce a set of simple gating functions $\{g_k\}_{k=1}^K$, where g_k maps the input to a scalar ranging in [0, 1] and $\sum_k g_k(\mathbf{x}) = 1$. Each sample in the mini-batch $\mathcal{B} = \{\mathbf{x}^{(i)}\}_{i=1}^B$ is then normalized under voting from its gate assignment:

$$ModeNorm(\mathbf{x}^{(i)}) = \gamma \left(\sum_{k=1}^K g_k(\mathbf{x}^{(i)}) \frac{\mathbf{x}^{(i)} - \mu_k}{\sigma_k} \right) + \beta, \; i = 1, \ldots, B, \quad (5.4)$$

where γ and β are the learned affine parameters, just as in standard BN. The estimators for mean μ_k and variance σ_k are computed under weighing from the gating network, $e.g.$, the k-th mean is estimated from the batch as:

$$\mu_k = \frac{1}{N_k} \sum_{\mathbf{x}^{(j)}\in\mathcal{B}} g_k(\mathbf{x}^{(j)}) \cdot \mathbf{x}^{(j)}, \quad (5.5)$$

where $N_k = \sum_{\mathbf{x}^{(j)}\in\mathcal{B}} g_k(\mathbf{x}^{(j)})$. The gate functions are trained jointly by backpropagation, which is different to MixNorm.

5.2   Combination

Since different normalization strategies have different advantages and disadvantages for training DNNs, some methods try to combine them. Luo et al. [5] proposed switchable normalization (SN), which combines three types of statistics, estimated channel-wise, layer-wise, and mini-batch-wise, by using IN, LN, and BN, respectively. SN switches between the different normalization methods by learning their importance weights, computed by a softmax function. SN has the expression as:

$$\hat{\mathbf{x}} = \gamma \frac{\mathbf{x} - \sum_{k\in\Omega} w_k \mu_k}{\sqrt{\sum_{k\in\Omega} w_k' \sigma_k^2 + \epsilon}} + \beta \quad (5.6)$$

where $\Omega = \{in, ln, bn\}$ is a set of statistics estimated in different NAPs. Furthermore, w_k and w'_k are importance ratios used to weighted average the means and variance respectively, ensuring $\sum_{k \in \Omega} w_k = 1$, $\sum_{k \in \Omega} w'_k = 1$ and $\forall w_k, w'_k \in [0, 1]$. The re-parameterization technique is used to calculate w_k (w'_k) by using a softmax function with $\{\alpha_{in}, \alpha_{ln}, \alpha_{bn}\}$ ($\{\alpha'_{in}, \alpha'_{ln}, \alpha'_{bn}\}$) as the control parameters

$$w_k = \frac{e^{\alpha_k}}{\sum_{z \in \{in, ln, bn\}} e^{\alpha_z}}, w'_k = \frac{e^{\alpha'_k}}{\sum_{z \in \{in, ln, bn\}} e^{\alpha'_z}}, and \; k \in \{in, ln, bn\}. \qquad (5.7)$$

$\{\alpha_{in}, \alpha_{ln}, \alpha_{bn}\}$ and $\{\alpha'_{in}, \alpha'_{ln}, \alpha'_{bn}\}$ can be learned by back-propagation. SN was designed to address the learning-to-normalize problem and obtains good results on several visual benchmarks [5]. Shao et al. [6] further introduced sparse switchable normalization (SSN), which selects different normalizations using the proposed SparsestMax function, which is a sparse version of softmax. Pan et al. [7] proposed switchable whitening (SW), which provides a general way to switch between different whitening and standardization methods under the SN framework. Zhang et al. [8] introduced exemplar normalization (EN) to investigate a dynamic 'learning-to-normalize' problem. Particularly, given a mini-batch input $\{\mathbf{x}^{(i)}\}_{i=1}^{B}$, EN is defined by

$$\hat{\mathbf{x}}^{(i)} = \sum_k \left(\gamma_k \left(w_k^{(i)} \frac{\mathbf{x}^{(i)} - \mu_k}{\sqrt{\sigma_k^2 + \epsilon}} \right) + \beta_k \right), \; i = 1, \ldots, B \qquad (5.8)$$

where $w_k^{(i)} \in [0, 1]$ indicates the important ratio of the k-th normalizer for the i-th sample, ensuring $\sum_k w_k^{(i)} = 1$. Furthermore the important ratios $w_k^{(i)}$ depends on the feature map of individual sample and is parameterized by sub-network with a self-attention structure. EN learns different data-dependent normalizations for different image samples, while SN fixes the importance ratios for the entire dataset. Gao et al. [9] propose representative batch normalization (RBN), which also combines the mini-batch statistics and instance-specific statistics for visual data. RBN utilizes instance-specific statistics to calibrate the centering and scaling operations with a negligible cost, and reduce the side effect introduced by some inappropriate running statistics while maintaining the benefits of BN. Besides, Luo et al. [10] proposed dynamic normalization (DN), which generalizes IN, LN, GN and BN in a unified formulation and can interpolate them to produce new normalization methods.

Considering that IN can learn style-invariant features [11], Nam and Kim [12] introduced batch-instance normalization (BIN) to normalize the styles adaptively to the task and selectively to individual feature maps. It learns to control how much of the style information is propagated through each channel by leveraging a learnable gate parameter $\rho \in [0, 1]^d$ to balance between IN and BN, as:

$$\hat{\mathbf{x}} = (\rho \cdot \hat{\mathbf{x}}_{BN} + (1 - \rho) \cdot \hat{\mathbf{x}}_{IN}) \cdot \gamma + \beta, \qquad (5.9)$$

where $\gamma, \beta \in \mathbb{R}^d$ are the affine parameters and $\hat{\mathbf{x}}_{BN}$ ($\hat{\mathbf{x}}_{IN}$) is the normalized output by BN (IN). The elements in ρ are constrained in the range [0, 1] by imposing bounds at the parameter update step:

$$\rho \leftarrow clip_{[0,1]}(\rho - \eta \Delta \rho), \tag{5.10}$$

where η is the learning rate and $\Delta \rho$ indicates the gradients with respect to ρ. A similar idea was also used in the adaptive layer-instance normalization (AdaLIN) [13] for image-to-image translation tasks, where a learnable gate parameter is leveraged to balance between LN and IN. Bronskill et al. [14] introduced TaskNorm, which combines LN/IN with BN for meta-learning scenarios. Rather than designing a combinational normalization module, Pan et al. [15] proposed IBN-Net, which carefully integrates IN and BN as building blocks, and can be wrapped into several deep networks to improve their performances. Qiao et al. [16] introduced batch-channel normalization (BCN), which integrates BN and channel-based normalizations (e.g., LN and GN) sequentially as a wrapped module.

Liu et al. [17] searched for a combination of normalization-activation layers using AutoML [18], leading to the discovery of EvoNorms, a set of new normalization-activation layers with sometimes surprising structures that go beyond existing design patterns.

References

1. Kalayeh, M. M. and M. Shah (2019). Training faster by separating modes of variation in batch-normalized models. *IEEE Transactions on Pattern Analysis and Machine Intelligence*.
2. Dempster, A. P., N. M. Laird, and D. B. Rubin (1977). Maximum likelihood from incomplete data via the em algorithm. *JOURNAL OF THE ROYAL STATISTICAL SOCIETY, SERIES B 39*(1), 1–38.
3. Arthur, D. and S. Vassilvitskii (2007). K-means++: The advantages of careful seeding. In *Proceedings of the Eighteenth Annual ACM-SIAM Symposium on Discrete Algorithms*.
4. Deecke, L., I. Murray, and H. Bilen (2019). Mode normalization. In *ICLR*.
5. Luo, P., J. Ren, Z. Peng, R. Zhang, and J. Li (2019). Differentiable learning-to-normalize via switchable normalization. In *ICLR*.
6. Shao, W., T. Meng, J. Li, R. Zhang, Y. Li, X. Wang, and P. Luo (2019). Ssn: Learning sparse switchable normalization via sparsestmax. In *CVPR*.
7. Pan, X., X. Zhan, J. Shi, X. Tang, and P. Luo (2019). Switchable whitening for deep representation learning. In *ICCV*.
8. Zhang, R., Z. Peng, L. Wu, Z. Li, and P. Luo (2020). Exemplar normalization for learning deep representation. In *CVPR*.
9. Gao, S.-H., Q. Han, D. Li, M.-M. Cheng, and P. Peng (2021, June). Representative batch normalization with feature calibration. In *Proceedings of the IEEE/CVF Conference on Computer Vision and Pattern Recognition (CVPR)*, pp. 8669–8679.
10. Luo, P., P. Zhanglin, S. Wenqi, Z. Ruimao, R. Jiamin, and W. Lingyun (2019). Differentiable dynamic normalization for learning deep representation. In *ICML*, pp. 4203–4211.
11. Ulyanov, D., A. Vedaldi, and V. S. Lempitsky (2016). Instance normalization: The missing ingredient for fast stylization. *arXiv preprint* arXiv:1607.08022.
12. Nam, H. and H.-E. Kim (2018). Batch-instance normalization for adaptively style-invariant neural networks. In *NeurIPS*.

13. Kim, J., M. Kim, H. Kang, and K. H. Lee (2020). U-gat-it: Unsupervised generative attentional networks with adaptive layer-instance normalization for image-to-image translation. In *ICLR*.
14. Bronskill, J., J. Gordon, J. Requeima, S. Nowozin, and R. E. Turner (2020b). Tasknorm: Rethinking batch normalization for meta-learning. In *ICML*.
15. Pan, X., P. Luo, J. Shi, and X. Tang (2018). Two at once: Enhancing learning and generalization capacities via IBN-net. In *ECCV*, pp. 484–500.
16. Qiao, S., H. Wang, C. Liu, W. Shen, and A. Yuille (2019a). Rethinking normalization and elimination singularity in neural networks. *arXiv preprint* arXiv:1911.09738.
17. Liu, H., A. Brock, K. Simonyan, and Q. V. Le (2020). Evolving normalization-activation layers. *arXiv preprint* arXiv:2004.02967.
18. Baker, B., O. Gupta, N. Naik, and R. Raskar (2017). Designing neural network architectures using reinforcement learning. In *ICLR*.

BN for More Robust Estimation

As illustrated in previous sections, BN introduces inconsistent normalization operations during training (using mini-batch statistics, as shown in Eq. 3.5) and inference (using population statistics estimated in Eq. 3.6). This means that the upper layers are trained on representations different from those computed during inference. These differences become significant if the batch size is too small, since the estimates of the mean and variance become less accurate. This leads to significantly degenerated performance [1–4]. To address this problem, some normalization methods avoid normalizing along the batch dimension, as introduced in previous sections. Here, we will discuss the more robust estimation methods that also address this problem of BN.

6.1 Normalization as Functions Combining Population Statistics

One way to reduce the discrepancy between training and inference is to combine the estimated population statistics for normalization during training.

Ioffe [1] proposed batch renormalization (BReNorm), augmenting the normalized output for each neuron with an affine transform, as:

$$\hat{x} = \frac{x - \mu_{\mathcal{B}}}{\sigma_{\mathcal{B}}} \cdot r + z, \tag{6.1}$$

where $r = \frac{\sigma_{\mathcal{B}}}{\hat{\sigma}}$ and $z = \frac{\mu_{\mathcal{B}} - \hat{\mu}}{\hat{\sigma}}$. Note that r and z are bounded between $\left(\frac{1}{r_{max}}, r_{max}\right)$ and $\left(\frac{1}{z_{max}}, z_{max}\right)$, respectively. Besides, r and z are treated as constants when performing gradient computation. Equation 6.1 is reduced to standardizing the activation using the estimated population (which ensures that the training and inference are consistent) if r and z are between their bounded values. Otherwise, Eq. 6.1 implicitly exploits the benefits of mini-batch statistics.

L. Huang, *Normalization Techniques in Deep Learning*, Synthesis Lectures on Computer Vision, https://doi.org/10.1007/978-3-031-14595-7_6

Dinh et al. [5] were the first to experiment with batch normalization using population statistics, which were weighted averages of the old population statistics and current mini-batch statistics, as shown in Eq. 3.6. The experimental results demonstrate that, combining population and mini-batch statistics can improve the performance of BN in small-batch-size scenarios. This idea is also used in diminishing batch normalization [6], full normalization (FN) [7], online normalization [8], moving average batch normalization (MABN) [9], PowerNorm [10] and momentum batch normalization (MBN) [11]. One challenge for this type of method is how to calculate the gradients during backpropagation, since the population statistics are computed by all the previous mini-batches, and it is impossible to obtain their exact gradients [12]. One straightforward strategy is to view the population statistics as constant and only back-propagate through current mini-batches, as proposed in [5–7]. However, this may introduce training instability, as discussed in Sect. 3.1. Chiley et al. [8] proposed to compute the gradients by maintaining the property of BN during backpropagation. Yan et al. [9] and Shen et al. [10] proposed to view the backpropagation gradients as statistics to be estimated, and approximate these statistics by moving averages.

Rather than explicitly using the population statistics, Guo et al. [13] introduced memorized BN, which considers data information from multiple recent batches (or all batches in an extreme case) to produce more accurate and stable statistics. A similar idea is exploited in cross-iteration batch normalization [14], where the mean and variance of examples from recent iterations are approximated for the current network weights via a low-order Taylor polynomial. Besides, Wang et al. [15] proposed Kalman normalization, which treats all the layers in a network as a whole system, and estimates the statistics of a certain layer by considering the distributions of all its preceding layers, mimicking the merits of Kalman filtering. Another practical approach for relieving the small-batch-size issue of BN in engineering systems is the synchronized batch normalization [16–18], which performs a synchronized computation of BN statistics across GPUs (Cross-GPU BN) to obtain better statistics.

6.2   Robust Inference Methods for BN

Some works address the small-batch-size problem of BN by finely estimating corrected normalization statistics during inference only. This strategy does not affect the training scheme of the model.

In fact, even the original BN paper [19] recommended estimating the population statistics after the training has finished (Algorithm 2 in [19]), rather than using the estimation calculated by running average, as shown in Eq. 3.6. However, while this can benefit a model trained with a small batch size, where estimation is the main issue [19–21], it may lead to degenerated generalization when the batch size is moderate.

Singh and Shrivastava [3] analyzed how a small batch size hampers the estimation accuracies of BN when using running averages, and proposed EvalNorm, which optimizes the sample weight during inference to ensure that the activations produced by normalization are similar to those provided during training. A similar idea is also exploited in [22], where the sample weights are viewed as hyperparameters, which are optimized on a validation set.

Compared to estimating the BN statistics (population mean and standardization deviation), Huang et al. [23] showed that estimating the whitening matrix of BW is more challenging. They demonstrated that, in terms of estimating the population statistics of the whitening matrix, it is more stable to use the mini-batch covariance matrix indirectly (the whitening matrix can be calculated after training) than the mini-batch whitening matrix directly.

References

1. Ioffe, S. (2017). Batch renormalization: Towards reducing minibatch dependence in batch-normalized models. In *NeurIPS*.
2. Wu, Y. and K. He (2018). Group normalization. In *ECCV*.
3. Singh, S. and A. Shrivastava (2019). Evalnorm: Estimating batch normalization statistics for evaluation. In *ICCV*.
4. Kaku, A., S. Mohan, A. Parnandi, H. Schambra, and C. Fernandez-Granda (2020). Be like water: Robustness to extraneous variables via adaptive feature normalization. *arXiv preprint* arXiv:2002.04019.
5. Dinh, L., J. Sohl Dickstein, and S. Bengio (2017). Density estimation using real NVP. In *ICLR*.
6. Ma, Y. and D. Klabjan (2017). Convergence analysis of batch normalization for deep neural nets. *arXiv preprint* arXiv:1705.08011.
7. Lian, X. and J. Liu (2019). Revisit batch normalization: New understanding and refinement via composition optimization. In *AISTATS*.
8. Chiley, V., I. Sharapov, A. Kosson, U. Koster, R. Reece, S. Samaniego de la Fuente, V. Subbiah, and M. James (2019). Online normalization for training neural networks. In *NeurIPS*.
9. Yan, J., R. Wan, X. Zhang, W. Zhang, Y. Wei, and J. Sun (2020). Towards stabilizing batch statistics in backward propagation of batch normalization. In *ICLR*.
10. Shen, S., Z. Yao, A. Gholami, M. W. Mahoney, and K. Keutzer (2020). Powernorm: Rethinking batch normalization in transformers. In *ICML*.
11. Yong, H., J. Huang, D. Meng, X. Hua, and L. Zhang (2020). Momentum batch normalization for deep learning with small batch size. In *ECCV*.
12. Liao, Q., K. Kawaguchi, and T. Poggio (2016). Streaming normalization: Towards simpler and more biologically-plausible normalizations for online and recurrent learning. *arXiv preprint* arXiv:1610.06160.
13. Guo, Y., Q. Wu, C. Deng, J. Chen, and M. Tan (2018). Double forward propagation for memorized batch normalization. In *AAAI*.
14. Yao, Z., Y. Cao, S. Zheng, G. Huang, and S. Lin (2021). Cross-iteration batch normalization. In *CVPR*.
15. Wang, G., J. Peng, P. Luo, X. Wang, and L. Lin (2018). Kalman normalization: Normalizing internal representations across network layers. In *NeurIPS*.
16. Zhao, H., J. Shi, X. Qi, X. Wang, and J. Jia (2017). Pyramid scene parsing network. In *CVPR*.
17. Liu, S., L. Qi, H. Qin, J. Shi, and J. Jia (2018). Path aggregation network for instance segmentation. In *CVPR*.
18. Peng, C., T. Xiao, Z. Li, Y. Jiang, X. Zhang, K. Jia, G. Yu, and J. Sun (2018). Megdet: A large mini-batch object detector. In *CVPR*.
19. Ioffe, S. and C. Szegedy (2015). Batch normalization: Accelerating deep network training by reducing internal covariate shift. In *ICML*.

20. Izmailov, P., D. Podoprikhin, T. Garipov, D. Vetrov, and A. G. Wilson (2018). Averaging weights leads to wider optima and better generalization. *arXiv preprint* arXiv:1803.05407.
21. Luo, P., J. Ren, Z. Peng, R. Zhang, and J. Li (2019). Differentiable learning-to-normalize via switchable normalization. In *ICLR*.
22. Summers, C. and M. J. Dinneen (2020). Four things everyone should know to improve batch normalization. In *ICLR*.
23. Huang, L., L. Zhao, Y. Zhou, F. Zhu, L. Liu, and L. Shao (2020). An investigation into the stochasticity of batch whitening. In *CVPR*.

Normalizing Weights

7

As stated in Chap. 2, normalizing the weights can implicitly normalize the activations by imposing constraints on the weight matrix, which can contribute to preserving the activations (gradients) during forward (backpropagation). Several seminal works have analyzed the distributions of the activations, given normalized inputs, under the assumption that the weights have certain properties or are under certain constraints, e.g., normalization propagation [1], variance propagation [2], self normalization [3], bidirectional self-normalization [4]. Specifically, in [1], a data independent estimate of the mean and variance statistics are available in closed form for every hidden layer, assuming the pre-activation values follow Gaussian distribution and that the weight matrix of hidden layers are roughly incoherent, as stated by following theories:

Theorem 7.1 (Canonical Error Bound [1]) *Consider a linear layer* $\mathbf{h} = \boldsymbol{W}\mathbf{x}$ *where* $\mathbf{x} \in \mathbb{R}^{d_{in}}$ *and* $\boldsymbol{W} \in \mathbb{R}^{d_{out} \times d_{in}}$ *such that* $\mathbb{E}_{\mathbf{x}}[\mathbf{x}] = \mathbf{0}$ *and* $\mathbb{E}_{\mathbf{x}}[\mathbf{x}\mathbf{x}^T] = \sigma^2 \boldsymbol{I}$. *Then the covariance matrix of* $\mathbf{h}$ *is approximately canonical satisfying*

$$\min_{\alpha} \| \Sigma - diag(\alpha) \|_F \leq \sigma^2 \tau \sqrt{\sum_{i,j=1; i \neq j}^{d_{out}} \| W_{i:} \|_2^2 \| W_{j:} \|_2^2}, \tag{7.1}$$

where $\Sigma = \mathbb{E}_{\mathbf{h}}[(\mathbf{h} - \mathbb{E}_{\mathbf{h}}[\mathbf{h}])(\mathbf{h} - \mathbb{E}_{\mathbf{h}}[\mathbf{h}])^T]$ *is the covariance matrix of* $\mathbf{h}$, τ *is the coherence*[1] *of the rows of* $\boldsymbol{W}$, $\alpha \in \mathbb{R}^{d_{out}}$ *is the closest approximation of the covariance matrix to a canonical ellipsoid and* $diag(\cdot)$ *diagonalizes a vector to a diagonal matrix. The corresponding optimal* $\alpha_i^* = \sigma^2 \| W_{i:} \|_2^2, \forall i \in \{1, \ldots, d_{out}\}$

---

[1] Coherence is defined as $\max_{W_{i:}, W_{j:}, i \neq j} \frac{|W_{i:}^T W_{j:}|}{\| W_{i:} \|_2 \| W_{j:} \|_2}$.

L. Huang, *Normalization Techniques in Deep Learning*, Synthesis Lectures on Computer Vision, https://doi.org/10.1007/978-3-031-14595-7_7

Theorem 7.2 (Post-ReLU distribution [1]) *Let* $h \sim \mathcal{N}(0, 1)$ *and* $z = max(0, h)$. *Then* $\mathbb{E}[z] = \frac{1}{\sqrt{2\pi}}$ *and* $var(z) = \frac{1}{2}(1 - \frac{1}{\pi})$.

Please refer to [1] for the proof of above theories. Theorems 7.1 and 7.2 shows how the mean and variance of activation can be analytically calculated through linear and non-linear (ReLU) transformations, given appropriate assumptions. We can see the key assumption is the properties imposed on the weight matrix W. The general idea of weight normalization is to provide layer-wise constraints on the weights during optimization, which can be formulated as:

$$\theta^* = \arg\min_\theta \mathbb{E}_{(\mathbf{x}, \mathbf{y}) \in D}[\mathcal{L}(\mathbf{y}, f(\mathbf{x}; \theta))]$$
$$s.t. \quad \Upsilon(W), \tag{7.2}$$

where $\Upsilon(W)$ are the layer-wise constraints imposed on the weights $W \in \mathbb{R}^{d_{out} \times d_{in}}$. It has been shown that the imposed constraints can benefit generalization [5–7]. In the following sections, we will introduce different constraints and discuss how to train a model with the constraints satisfied.

7.1   Constraints on Weights

Reference [8] proposed weight normalization (WN), which requires the input weight of each neuron to be unit norm. Specifically, given one neuron's input weight $W_{i:} \in \mathbb{R}^{d_{in}}$, the constraints imposed on W are:

$$\Upsilon(W) = \{\|W_{i:}\| = 1, i = 1, \ldots, d_{out}\}. \tag{7.3}$$

We can found the error bound shown in Eq. 7.1 will be reduced to $\sigma^2 \tau \sqrt{d_{out}(d_{out} - 1)}$ and the corresponding optimal $\alpha_i^* = \sigma^2$, if we impose the constraints on the weight by Eq. 7.3. In this case, We may preserve the variance among linear transformations and the error bound only depends on the coherence τ. WN has a scale-invariant property like BN, which is important for stabilizing training.

Inspired by the practical weight initialization technique [9, 10], where weights are sampled from a distribution with zero mean and a standard deviation for initialization, Huang et al. [5] further proposed centered weight normalization (CWN), constraining the input weight of each neuron to have zero mean and unit norm, as:

$$\Upsilon(W) = \{W_{i:}^T \mathbf{1} = 0 \,\&\, \|W_{i:}\| = 1, i = 1, \ldots, d_{out}\}. \tag{7.4}$$

CWN can theoretically preserve the activation statistics (mean and variance) between different layers under certain assumptions, which can benefit optimization, as stated by following theory.

Theorem 7.3 *Let consider a neuron* $h = \mathbf{w}^T\mathbf{x}$, *where* $\mathbf{w}^T\mathbf{1} = 0$ *and* $\|\mathbf{w}\| = 1$. *Assume* $\mathbf{x}$ *has Gaussian distribution with the mean:* $\mathbb{E}_\mathbf{x}[\mathbf{x}] = \mu\mathbf{1}$, *and covariance matrix:* $cov(\mathbf{x}) = \sigma^2\mathbf{I}$, *where* $\mu \in \mathbb{R}$ *and* $\sigma^2 \in \mathbb{R}$. *We have* $\mathbb{E}_h[h] = 0$, $var(h) = \sigma^2$.

Such a theory tells us that for each neuron the pre-activation h has zero-mean and the same variance as the activations fed in, when the assumption is satisfied. Weight centering is also advocated for in [11, 12]. Qiao et al. proposed weight standardization (WS), which imposes the constraints on the weights with $\Upsilon(W) = \{W_{i:}^T\mathbf{1} = 0 \ \& \ \|W_{i:}\| = \sqrt{d_{out}}, i = 1, \dots, d_{out}\}$. Note that WS cannot effectively preserve the activation statistics between different layers, since the weight norm is $\sqrt{d_{out}}$, which may cause exploding activations. Therefore, WS usually needs to be combined with activation normalization methods (e.g., BN/GN) to relieve this issue [11].

Another widely used constraint on weights is orthogonality, which is represented as

$$\Upsilon(W) = \{WW^T = I\}. \tag{7.5}$$

Given the orthogonal constraints shown in Eq. 7.5, We can found the error bound shown in Eq. 7.1 will be zero (the coherence τ will be zero) and the corresponding optimal $\alpha_i^* = \sigma^2$, i.e., the covariance matrix of $\mathbf{h}$ will be isometric. Following theory shows the advantages of the orthogonality constraints in preserving the norm and distribution of the activation for a linear transformation.

Theorem 7.4 *Let* $\mathbf{h} = W\mathbf{x}$, *where* $WW^T = I$ *and* $W \in \mathbb{R}^{d_{out} \times d_{in}}$. *Assume: (1)* $\mathbb{E}_\mathbf{x}(\mathbf{x}) = \mathbf{0}$, $cov(\mathbf{x}) = \sigma_1^2\mathbf{I}$, *and (2)* $\mathbb{E}_{\frac{\partial \mathcal{L}}{\partial \mathbf{h}}}(\frac{\partial \mathcal{L}}{\partial \mathbf{h}}) = \mathbf{0}$, $cov(\frac{\partial \mathcal{L}}{\partial \mathbf{h}}) = \sigma_2^2\mathbf{I}$. *If* $d_{out} = d_{in}$, *we have the following properties: (1)* $\|\mathbf{h}\| = \|\mathbf{x}\|$; *(2)* $\mathbb{E}_\mathbf{h}(\mathbf{h}) = \mathbf{0}$, $cov(\mathbf{h}) = \sigma_1^2\mathbf{I}$; *(3)* $\|\frac{\partial \mathcal{L}}{\partial \mathbf{x}}\| = \|\frac{\partial \mathcal{L}}{\partial \mathbf{h}}\|$; *(4)* $\mathbb{E}_{\frac{\partial \mathcal{L}}{\partial \mathbf{x}}}(\frac{\partial \mathcal{L}}{\partial \mathbf{x}}) = \mathbf{0}$, $cov(\frac{\partial \mathcal{L}}{\partial \mathbf{x}}) = \sigma_2^2\mathbf{I}$. *In particular, if* $d_{out} < d_{in}$, *property (2) and (3) hold; if* $d_{out} > d_{in}$, *property (1) and (4) hold.*

Orthogonality was first used in the square hidden-to-hidden weight matrices of RNNs [13–19], and then further extended to the more general rectangular matrices in DNNs [6, 20–23]. Orthogonal weight matrices can theoretically preserve the norm of activations/output-gradients between linear transformations [22–24]. Further, the distributions of activations/output-gradients can also be preserved under mild assumptions [6, 24]. These properties of orthogonal weight matrices are beneficial for the optimization of DNNs. Furthermore, orthogonal weight matrices can avoid learning redundant filters, benefitting generalization.

Rather than bounding all singular values as 1, like in orthogonal weight matrices, Miyato et al. [25] proposed spectral normalization, which constrains the spectral norm (the maximum singular value) of a weight matrix to 1, in order to control the Lipschitz constant of the discriminator when training GANs. Huang et al. [24] proposed orthogonalization by Newton's iterations (ONI), which controls the orthogonality through the iteration number.

They showed that it is possible to bound the singular values of a weight matrix between $(\sigma_{min}, 1)$ during training. ONI effectively interpolates between spectral normalization and full orthogonalization, by altering the iteration number.

Note that the constraints imposed on the weight matrix (Eqs. 7.3, 7.4, 7.5) may harm the representation capacity and result in degenerated performance. An extra learnable scalar parameter is usually used for each neuron to recover the possible loss in representation capacity, which is similar to the idea of the affine parameters proposed in BN.

7.2   Training with Constraints

It is clear that training a DNN with constraints imposed on the weights is a constraint optimization problem. Here, we summarize three kinds of strategies for solving this.

Re-Parameterization. One stable way to solve constraint optimization problems is to use a re-parameterization method (Fig. 7.1). Re-parameterization constructs a fine transformation ψ over the proxy parameter V to ensure that the transformed weight W has certain beneficial properties for the training of neural networks. Gradient updating is executed on the proxy parameter V by back-propagating the gradient information through the normalization

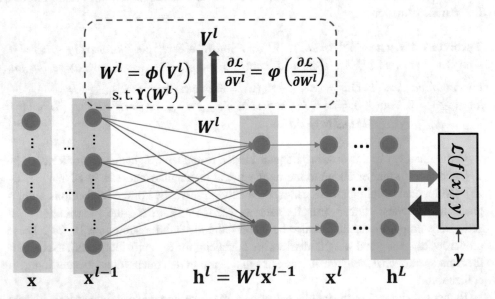

Fig. 7.1 An illustrative example of neural networks with layer-wised re-parameterization. Re-parameterization constructs a fine transformation ψ over the proxy parameter V to ensure that the transformed weight W has certain beneficial properties for the training of neural networks. Gradient updating is executed on the proxy parameter V by back-propagating the gradient information through the normalization process

process. Re-parameterization was first used in learning the square orthogonal weight matrices in RNNs [13–15]. Salimans and Kingma [8] used this technique to learn a unit-norm constraint as shown in Eq. 7.3. Huang et al. [5] formally described the re-parameterization idea in training with constraints on weight matrices, and applied it to solve the optimization with zero-mean and unit-norm constraints as shown in Eq. 7.4. This technique was then further used in other methods for learning with different constraints, e.g., orthogonal weight normalization [6], weight standardization [11], spectral normalization [25] and weight centralization [12]. Re-parameterization is a main technique for optimizing constrained weights in DNNs. Its main merit is that the training is relatively stable, because it updates V based on the gradients computed by backpropagation, while simultaneously maintaining the constraints on W. The downside is that the backpropagation through the designed transformation may increase the computational cost.

Regularization with an Extra Penalty. Some works have tried to maintain the weight constraints using an additional penalty on the objective function, which can be viewed as a regularization. This regularization technique is mainly used for learning the weight matrices with orthogonality constraints, for its efficiency in computation [16, 26–28]. Orthogonal regularization methods have demonstrated improved performance in image classification [27, 29, 30], resisting attacks from adversarial examples [31], neural photo editing [32] and training GANs [25, 33]. However, the introduced penalty works like a pure regularization, and whether or not the constraints are truly maintained or training benefited is unclear. Besides, orthogonal regularization usually requires to be combined with activation normalization, when applied on large-scale architectures, since it cannot stabilize training.

Riemannian Optimization. A weight matrix W with constraints can be viewed as an embedded submanifold [6, 34–36]. For example, a weight matrix with an orthogonality constraint (Eq. 7.5) is a real Stiefel manifold [6, 34]. One possible way of maintaining these constraints when training DNNs is to use Riemannian optimization [37, 38]. Here, we review the Riemannian optimization over Stiefel manifold briefly and for more details please refer to [39] and references therein. The objective is $\arg\min_{W \in \mathbb{M}} f(W)$, where f is a real-value smooth function over $\mathbb{M} = \{W \in \mathbb{R}^{n \times d} : W^T W = I, n \geqslant d\}$. We follow the common description for Riemannian optimization over Stiefel manifold with the columns of W being d orthonormal vectors in $\mathbb{R}^n$, and therefore with the constraints $W^T W = I$. It is different to the description of our formulation in the previous sections with constraints $WW^T = I$ and $n \leqslant d$, shown in Eq. 7.5.

Conventional optimization techniques are based gradient descent method over manifold by iteratively seeking for updated points $W_t \in \mathbb{M}$. In each iteration t, the keys are: (1) finding the Riemannian gradient $G^{\mathbb{M}} f(W_t) \in T_{W_t}$ where T_{W_t} is the tangent space of $\mathbb{M}$ at current point W_t; and (2) finding the descent direction and ensuring that the new points is on the manifold $\mathbb{M}$.

For obtaining the Riemannian gradient $G^{\mathbb{M}} f(W)$, the inner dot should be defined in T_W. There are two extensively used inner products for tangent space of Stiefel manifold [40]: (1) *Euclidean inner product*: $< \mathbf{X}_1, \mathbf{X}_2 >_e = tr(\mathbf{X}_1^T \mathbf{X}_2)$ and (2) *canonical inner product*:

$< \mathbf{X}_1, \mathbf{X}_2 >_c = tr(\mathbf{X}_1^T (\mathbf{I} - \frac{1}{2}\mathbf{W}\mathbf{W}^T)\mathbf{X}_2)$ where $\mathbf{X}_1, \mathbf{X}_2 \in T_W$ and $tr(\cdot)$ denote the trace of the matrix. Based on these two inner products, the respective Riemannian gradient can be obtained as [40]:

$$G_e^{\mathbb{M}} f(\mathbf{W}) = \frac{\partial f}{\partial \mathbf{W}} - \mathbf{W} \frac{\partial f}{\partial \mathbf{W}}^T \mathbf{W} \qquad (7.6)$$

and

$$G_c^{\mathbb{M}} f(\mathbf{W}) = \frac{\partial f}{\partial \mathbf{W}} - \frac{1}{2}\left(\mathbf{W}\mathbf{W}^T \frac{\partial f}{\partial \mathbf{W}} + \mathbf{W} \frac{\partial f}{\partial \mathbf{W}}^T \mathbf{W} \right) \qquad (7.7)$$

where $\frac{\partial f}{\partial \mathbf{W}}$ is the ordinary gradient.

Given the Riemannian gradient, the next step is to find the descent direction and guarantee that the new point is on the manifold $\mathbb{M}$, which can be supported by the so called operation *retraction*. One well recommended *retraction* is the QR-Decomposition-Type retraction [38, 41] that maps a tangent vector of T_W onto $\mathbb{M}$ by: $P_W(\mathbf{Z}) = qf(\mathbf{W} + \mathbf{Z})$, where $qf(\cdot)$ denotes the Q factor of the QR decomposition with $Q \in \mathbb{M}$, and the R-factor is an upper-trangular matrix with strictly positive elements on its main diagonal such that the decomposition is unique [41]. Given the Riemannian gradient $G^{\mathbb{M}} f(\mathbf{W})$ and the learning rate η, the new point is:

$$\mathbf{W}_{t+1} = qf(\mathbf{W}_t - \eta\, G^{\mathbb{M}} f(\mathbf{W})) \qquad (7.8)$$

Another well known technique to jointly move in the descent direction and make sure the new solution on the manifold $\mathbb{M}$ is Cayley transformation [14, 16, 36, 40]. It produces the feasible solution $\mathbf{W}_{t+1}$ with the current solution $\mathbf{W}_t$ by:

$$\mathbf{W}_{t+1} = \left(\mathbf{I} + \frac{\eta}{2}A_t \right)^{-1} \left(\mathbf{I} - \frac{\eta}{2}A_t \right) \mathbf{W}_t \qquad (7.9)$$

where η is the learning rate and $A_t = \frac{\partial f}{\partial \mathbf{W}_t}^T \mathbf{W}_t - \mathbf{W}_t^T \frac{\partial f}{\partial \mathbf{W}_t}$ that is induced by the defined canonical inner product in the tangent space.

The main difficulties of applying Riemannian optimization in training DNNs are: (1) The optimization space covers multiple embedded submanifolds; (2) The embedded submanifolds are inter-dependent since the optimization of the current weight layer is affected by those of preceding layers. To stabilize the training, activation normalizations (e.g., BN) [36] or gradient clips [34] are usually required. One interesting observation is that using BN will probably improve the performance of projection-based methods (where the gradient is calculated based on the Euclidean space) [35, 42].

Here, we summarize the advantages and disadvantages of normalizing weights, compared to normalizing activations. The main advantages of normalizing weights are their efficiency during inference, i.e., there are no additional memory or computation cost during inference.

Besides, they are not sensitive to the batch size, compared to batch normalization. The main disadvantage of normalizing weights is that the training is potentially not stable, compared to activation normalizations. Because they can only implicitly control the activations and usually requires certain assumptions. Besides, they are required to well design the gain parameters for maintaining the equivalent variance/distribution among layers [24, 43]. The gain parameters depend on the network architectures. For example, Huang et al. [24] consider the feed forward network without residual connections, and Brock et al. [43] consider the more difficult residual network architectures. Therefore, they are more difficult to use in practice.

References

1. Arpit, D., Y. Zhou, B. U. Kota, and V. Govindaraju (2016). Normalization propagation: A parametric technique for removing internal covariate shift in deep networks. In *ICML*.
2. Shekhovtsov, A. and B. Flach (2018a). Normalization of neural networks using analytic variance propagation. In *Computer Vision Winter Workshop*.
3. Klambauer, G., T. Unterthiner, A. Mayr, and S. Hochreiter (2017). Self-normalizing neural networks. In *NeurIPS*.
4. Lu, Y., S. Gould, and T. Ajanthan (2020). Bidirectional self-normalizing neural networks. arXiv preprint arXiv:2006.12169.
5. Huang, L., X. Liu, Y. Liu, B. Lang, and D. Tao (2017). Centered weight normalization in accelerating training of deep neural networks. In *ICCV*.
6. Huang, L., X. Liu, B. Lang, A. W. Yu, Y. Wang, and B. Li (2018). Orthogonal weight normalization: Solution to optimization over multiple dependent stiefel manifolds in deep neural networks. In *AAAI*.
7. Özay, M. and T. Okatani (2018). Training CNNs with normalized kernels. In *AAAI*.
8. Salimans, T. and D. P. Kingma (2016). Weight normalization: A simple reparameterization to accelerate training of deep neural networks. In *NeurIPS*.
9. Glorot, X. and Y. Bengio (2010). Understanding the difficulty of training deep feedforward neural networks. In *AISTATS*.
10. He, K., X. Zhang, S. Ren, and J. Sun (2015). Delving deep into rectifiers: Surpassing human-level performance on imagenet classification. In *ICCV*.
11. Qiao, S., H. Wang, C. Liu, W. Shen, and A. Yuille (2019b). Weight standardization. arXiv preprint arXiv:1903.10520.
12. Yan, J., R. Wan, X. Zhang, W. Zhang, Y. Wei, and J. Sun (2020). Towards stabilizing batch statistics in backward propagation of batch normalization. In *ICLR*.
13. Arjovsky, M., A. Shah, and Y. Bengio (2016). Unitary evolution recurrent neural networks. In *ICML*.
14. Wisdom, S., T. Powers, J. Hershey, J. Le Roux, and L. Atlas (2016). Full-capacity unitary recurrent neural networks. In *NeurIPS*.
15. Dorobantu, V., P. A. Stromhaug, and J. Renteria (2016). DizzyRNN: Reparameterizing recurrent neural networks for norm-preserving backpropagation. arXiv preprint arXiv:1612.04035.
16. Vorontsov, E., C. Trabelsi, S. Kadoury, and C. Pal (2017). On orthogonality and learning recurrent networks with long term dependencies. In *ICML*.
17. Hyland, S. and G.Rätsch (2017). Learning unitary operators with help from u(n). In *AAAI*.

18. Jing, L., Ç. Gülçehre, J. Peurifoy, Y. Shen, M. Tegmark, M. Soljacic, and Y. Bengio (2017). Gated orthogonal recurrent units: On learning to forget. arXiv preprint arXiv:1706.02761.
19. Helfrich, K., D. Willmott, and Q. Ye (2018). Orthogonal recurrent neural networks with scaled Cayley transform. In *ICML*.
20. Ozay, M. (2019). Fine-grained optimization of deep neural networks. In *NeurIPS*.
21. Jia, K., S. Li, Y. Wen, T. Liu, and D. Tao (2019). Orthogonal deep neural networks. *IEEE transactions on pattern analysis and machine intelligence*.
22. Wang, J., Y. Chen, R. Chakraborty, and S. X. Yu (2020). Orthogonal convolutional neural networks. In *CVPR*.
23. Qi, H., C. You, X. Wang, Y. Ma, and J. Malik (2020). Deep isometric learning for visual recognition. In *ICML*.
24. Huang, L., L. Liu, F. Zhu, D. Wan, Z. Yuan, B. Li, and L. Shao (2020). Controllable orthogonalization in training DNNs. In *CVPR*.
25. Miyato, T., T. Kataoka, M. Koyama, and Y. Yoshida (2018). Spectral normalization for generative adversarial networks. In *ICLR*.
26. Pascanu, R., T. Mikolov, and Y. Bengio (2013). On the difficulty of training recurrent neural networks. In *ICML*.
27. Bansal, N., X. Chen, and Z. Wang (2018). Can we gain more from orthogonality regularizations in training deep CNNs? In *NeurIPS*.
28. Amjad, J., Z. Lyu, and M. R. Rodrigues (2019). Deep learning for inverse problems: Bounds and regularizers. arXiv preprint arXiv:1901.11352.
29. Zhang, L., M. Edraki, and G.-J. Qi (2018). Cappronet: Deep feature learning via orthogonal projections onto capsule subspaces. In *NeurIPS*.
30. Lezama, J., Q. Qiu, P. Musé, and G. Sapiro (2018). OlÉ: Orthogonal low-rank embedding - a plug and play geometric loss for deep learning. In *CVPR*.
31. Moustapha, C., B. Piotr, E. Grave, Y. Dauphin, and N. Usunie (2017). Parseval networks: Improving robustness to adversarial examples. In *ICML*.
32. Brock, A., T. Lim, J. M. Ritchie, and N. Weston (2017). Neural photo editing with introspective adversarial networks. In *ICLR*.
33. Brock, A., J. Donahue, and K. Simonyan (2019). Large scale GAN training for high fidelity natural image synthesis. In *ICLR*.
34. Cho, M. and J. Lee (2017). Riemannian approach to batch normalization. In *NeurIPS*.
35. Huang, L., X. Liu, B. Lang, and B. Li (2017). Projection based weight normalization for deep neural networks. arXiv preprint arXiv:1710.02338.
36. Li, J., L. Fuxin, and S. Todorovic (2020). Efficient riemannian optimization on the stiefel manifold via the cayley transform. In *ICLR*.
37. Ozay, M. and T. Okatani (2016). Optimization on submanifolds of convolution kernels in CNNs. arXiv preprint arXiv:1610.07008.
38. Harandi, M. and B. Fernando (2016). Generalized backpropagation, etude de cas: Orthogonality. arXiv preprint arXiv:1611.05927.
39. Absil, P.-A., R. Mahony, and R. Sepulchre (2008). *Optimization Algorithms on Matrix Manifolds*. Princeton, NJ: Princeton University Press.
40. Wen, Z. and W. Yin (2013). A feasible method for optimization with orthogonality constraints. *Math. Program. 142*(1-2), 397–434.
41. Kaneko, T., S. G. O. Fiori, and T. Tanaka (2013). Empirical arithmetic averaging over the compact stiefel manifold. *IEEE Trans. Signal Processing 61*(4), 883–894.
42. Jia, K. (2017). Improving training of deep neural networks via singular value bounding. In *CVPR*.
43. Brock, A., S. De, and S. L. Smith (2021). Characterizing signal propagation to close the performance gap in unnormalized resnets. In *International Conference on Learning Representations*.

Normalizing Gradients

8

As stated previously, normalizing activations and weights aims to provide a better optimization landscape for DNNs, by satisfying Criteria 1 and 2 in Chap. 2. Rather than providing a good optimization landscape by design, normalizing gradients in DNNs aims to exploit the curvature information for GD/SGD, even though the optimization landscape is ill-conditioned. It performs normalization solely on the gradients, which may effectively remove the negative effects of an ill-conditioned landscape caused by the diversity in magnitude of gradients from different layers [1]. Generally speaking, normalizing gradients is similar to second-order optimization [2–5] or coordinate-wise adaptive learning rate based methods [6–8], but with the goal of exploiting the layer-wise structural information in DNNs.

Yu et al. [9] were the first to propose block-wise (layer-wise) gradient normalization for training DNNs to front the gradient explosion or vanishing problem. The generic block-normalized gradient descent are described in Algorithm 8.1.

Algorithm 8.1 Generic Block-Normalized Gradient (BNG) Descent

---

1: **Parameters:** number of steps T, number of blocks B

2: Divide θ into B blocks such that $\theta = (\theta^1, \theta^2, \ldots, \theta^B)$. Initialize $\theta_0 \in \mathbb{R}^D$.

3: **for** $t = 1, 2, \ldots T$ **do**

4:     Sample a mini-batch of data X_t and compute the normalized stochastic gradient w.r.t. each block

$$\mathbf{g}_t^i = \frac{\frac{\partial \mathcal{L}}{\partial \theta^i}}{\|\frac{\partial \mathcal{L}}{\partial \theta^i}\|_2}, i = 1, 2, \ldots, B$$

5:     Let $\mathbf{g}_t = (\mathbf{g}_t^1, \mathbf{g}_t^2, \ldots, \mathbf{g}_t^B)$ and choose step sizes $\tau_t \in \mathbb{R}^D$.

6:     $\theta_t = \theta_{t-1} - \tau_t \cdot \mathbf{g}_t$

7: **end for**

Specifically, they perform scaling over the gradients w.r.t. the weight in each layer, ensuring the norm to be unit-norm. I.e., the blocks described in Algorithm 8.1 corresponds to

L. Huang, *Normalization Techniques in Deep Learning*, Synthesis Lectures on Computer Vision, https://doi.org/10.1007/978-3-031-14595-7_8

the layers in a network. This technique can decrease the magnitude of a large gradient to a certain level, like gradient clipping [10], and also boost the magnitude of a small gradient. However, the net-gain of this approach degenerates in the scale-invariant DNNs (e.g., with BN). In [9], an extra ratio factor that depends on the norm of the layer-wise weight was used to adaptively adjust the magnitude of the gradients, as follows:

$$\mathbf{g}_{adap}^i = \alpha \cdot \|\theta_i\| \frac{\frac{\partial \mathcal{L}}{\partial \theta^i}}{\|\frac{\partial \mathcal{L}}{\partial \theta^i}\|_2}, i = 1, 2, \ldots, B. \tag{8.1}$$

Here α is a constant ratio and the subscript "adap" is short for "adaptive", as the resulting norm of the gradient is adaptive to its variable norm.

A similar idea was also concurrently introduced in the layer-wise adaptive rate scaling (LARS), proposed by You et al. [11], for large-batch training. LARS provides a detailed illustration on how to use weight decay, momentum in the normalizing gradient framework. The network training for SGD with LARS are summarized in the Algorithm 8.2.

Algorithm 8.2 SGD with LARS, using weight decay, momentum and polynomial learning rate (LR) decay

---

1: **Parameters:** base LR γ_0, momentum β_m, weight decay β, LARS coefficient η, number of steps T
2: **Init:** momentum vector $\mathbf{v}_0 = 0$, weight $\mathbf{w}_0^l$ for each layer l.
3: **for** $t = 1, 2, \ldots T$ **do**
4:     Compute the global LR: $\gamma_t = \gamma_0 * (1 - \frac{t}{T})^2$
5:     **for** each layer l **do**
6:         Sample a mini-batch of data and compute the mini-batch gradient: $\mathbf{g}_t^l = \frac{\partial \mathcal{L}}{\partial \mathbf{w}^l}|_{\mathbf{w}^l = \mathbf{w}_{t-1}^l}$
7:         Compute the local LR: $\lambda^l = \eta * \frac{\|\mathbf{w}_{t-1}^l\|}{\|\mathbf{g}_t^l\| + \beta \|\mathbf{w}_{t-1}^l\|}$
8:         Update the momentum: $\mathbf{v}_t^l = \beta_m \mathbf{v}_{t-1}^l + \gamma_t * \lambda^l * (\mathbf{g}_t^l + \beta \mathbf{w}_{t-1}^l)$
9:         Update the weights: $\mathbf{w}_t^l = \mathbf{w}_{t-1}^l - \mathbf{v}_t^l$
10:     **end for**
11: **end for**

LARS is an essential technique in training large-scale DNNs using large batch sizes, significantly reducing the training times without degradation of performance, especially for visual tasks (e.g., on ImageNet) using SGD. You et al. [12] further proposed a layer-wise adaptive large batch optimization technique called LAMB for training state-of-the-art language models like BERT. LAMB is based on the popular Adam optimizer [8] in deep learning community, and is described in Algorithm 8.3. Note that LAMB generalizes the layerwise scaling term as a scaling function $\phi(\|\mathbf{w}^l\|)$ whose value depends the norm of the weight vector $\mathbf{w}^l$ in layer l. LAMB is a prominent example that reduces the training time of BERT from 3 days to 76 min on a TPUv3 Pod. Zheng et al. [13] introduce per-block gradient normalization to LAMB and modify its momentum term by taking advantage of the connection between the classic momentum and Nesterov's momentum. The resultant accelerated gradient method is called LANS.

Algorithm 8.3 LAMB

1: **Parameters:** number of steps T, learning rate $\{\gamma_t\}_{t=1}^T$, momentum parameters $0 < \beta_1, \beta_2 < 1$, weight decay β, scaling functions ϕ, $\epsilon > 0$
2: **Init:** momentum vectors $\mathbf{m}_0 = \mathbf{0}$, $\mathbf{v}_0 = \mathbf{0}$, weight $\mathbf{w}_0^l$ for each layer l.
3: **for** $t = 0, 1, 2, \ldots T - 1$ **do**
4:     Sample a mini-batch of data and compute the mini-batch gradient: $\mathbf{g}_t = \frac{\partial \mathcal{L}}{\partial \mathbf{w}}|_{\mathbf{w}=\mathbf{w}_{t-1}}$
5:     $\mathbf{m}_t = \beta_1 \mathbf{m}_{t-1} + (1 - \beta_1)\mathbf{g}_t$
6:     $\mathbf{v}_t = \beta_2 \mathbf{v}_{t-1} + (1 - \beta_2)\mathbf{g}_t^2$
7:     $\mathbf{m}_t = \mathbf{m}_t/(1 - \beta_1^t)$
8:     $\mathbf{v}_t = \mathbf{v}_t/(1 - \beta_2^t)$
9:     Compute ratio $\mathbf{r}_t = \frac{\mathbf{m}_t}{\sqrt{\mathbf{v}_t}+\epsilon}$ (element-wise division)
10:    **for** each layer l **do**
11:        Update the weights: $\mathbf{w}_t^l = \mathbf{w}_{t-1}^l - \gamma_t \frac{\phi(\|\mathbf{w}_{t-1}^l\|)}{\|\mathbf{r}_t^l+\beta\mathbf{w}_{t-1}^l\|}(\mathbf{r}_t^l + \beta\mathbf{w}_{t-1}^l)$
12:    **end for**
13: **end for**

Rather than using the scaling operation, Yong et al. [14] recently proposed gradient centralization (GC), which performs centering over the gradient w.r.t. the input weight of each neuron in each layer. Specifically, given the weight matrix $W \in \mathbb{R}^{d_{out} \times d_{in}}$ in fully-connected layer where d_{out} indicates the number of neurons, GC centers the gradient as follows:

$$\phi_{GC}\left(\frac{\partial \mathcal{L}}{\partial W}\right) = \frac{\partial \mathcal{L}}{\partial W}\left(I - \frac{1}{d_{in}}\mathbf{1}\mathbf{1}^T\right). \tag{8.2}$$

The centered gradient $\phi_{GC}(\frac{\partial \mathcal{L}}{\partial W})$ is then used to update the weight. GC implicitly imposes constraints on the input weight, and ensures that the sum of elements in the input weight is a constant during training. GC effectively improves the performances of DNNs with activation normalization (e.g., BN or GN).

References

1. Glorot, X. and Y. Bengio (2010). Understanding the difficulty of training deep feedforward neural networks. In *AISTATS*.
2. Martens, J. (2010). Deep learning via Hessian-free optimization. In *ICML*, pp. 735–742. Omnipress.

3. Vinyals, O. and D. Povey (2012). Krylov subspace descent for deep learning. In *AISTATS*, pp. 1261–1268.

4. Martens, J. and I. Sutskever (2012). Training deep and recurrent networks with Hessian-free optimization. In *Neural Networks: Tricks of the Trade (2nd ed.)*, Volume 7700 of *Lecture Notes in Computer Science*, pp. 479–535. Springer.

5. Grosse, R. B. and R. Salakhutdinov (2015). Scaling up natural gradient by sparsely factorizing the inverse fisher matrix. In *ICML*, pp. 2304–2313.

6. Duchi, J., E. Hazan, and Y. Singer (2011). Adaptive subgradient methods for online learning and stochastic optimization. *Journal of machine learning research 12*(7).

7. Hinton, G., N. Srivastava, and K. Swersky (2012). Neural networks for machine learning lecture 6a overview of mini-batch gradient descent.

8. Kingma, D. P. and J. Ba (2015). Adam: A method for stochastic optimization. In *ICLR*.

9. Yu, A. W., L. Huang, Q. Lin, R. Salakhutdinov, and J. Carbonell (2017). Block-normalized gradient method: An empirical study for training deep neural network. arXiv preprint arXiv:1707.04822.

10. Pascanu, R., T. Mikolov, and Y. Bengio (2013). On the difficulty of training recurrent neural networks. In *ICML*.

11. You, Y., I. Gitman, and B. Ginsburg (2017). Large batch training of convolutional networks. arXiv preprint arXiv:1708.03888.

12. You, Y., J. Li, S. Reddi, J. Hseu, S. Kumar, S. Bhojanapalli, X. Song, J. Demmel, K. Keutzer, and C.-J. Hsieh (2020). Large batch optimization for deep learning: Training BERT in 76 minutes. In *ICLR*.

13. Zheng, S., H. Lin, S. Zha, and M. Li (2020). Accelerated large batch optimization of BERT pretraining in 54 minutes. *CoRR abs/2006.13484*.

14. Yong, H., J. Huang, X. Hua, and L. Zhang (2020). Gradient centralization: A new optimization technique for deep neural networks. In *ECCV*.

Analysis of Normalization

In Chap. 2, we provided high-level motivation of normalization in benefiting network optimization. In this section, we will further discuss other properties of normalization methods in improving DNNs' training performance. Regards to theoretic analyses of machine learning or deep learning, there usually involve three main topics: representation, optimization, and generalization. In terms of studying representation power in deep learning, it mainly covers to quantitatively measure the complexity of the deep models, and theoretically characterize how depth of DNNs benefits the representation capacity, compared to the width [1, 2]. Given a model, the optimization study mainly includes whether a stationary point can be obtained by a optimization algorithm or how fast the optimization algorithm to arrive the stationary point. Generalization refers to the capability of models well trained on the training data by optimization algorithms predicting on the unseen data.

In this chapter, we mainly focus on BN, because it displays nearly all the benefits of normalization in improving the performance of DNNs, e.g., stabilizing training, accelerating convergence and improving the generalization. We will show that how BN benefits optimization and generalization, and show how BN affects the representation by using its batch size to control the distribution. We address that there remains little consensus on the exact reason and mechanism behind BN's performance, the theoretic analyses for BN and its effectiveness in DNNs is still an active research topic. Here, we don't discuss rigorous theoretic analyses, in terms of answering what is the bound of access risk or generalization errors of normalized models, or whether BN can improve the theoretic convergence rate. It is difficult for the theoretical analyses of deep neural networks, even there is no normalization in a network. This book mainly focuses on illustrate the analyses that combines high-level theoretic intuition and the empirical experimental validation.

© The Author(s), under exclusive license to Springer Nature Switzerland AG 2022   65
L. Huang, *Normalization Techniques in Deep Learning*, Synthesis Lectures on Computer Vision, https://doi.org/10.1007/978-3-031-14595-7_9

9.1   Scale Invariance in Stabilizing Training

One essential functionality of BN is its ability to stabilize training. This is mainly due to its scale-invariant property [3–6], i.e., it does not change the prediction when rescaling parameters and works by adaptively adjusting the learning rate in a layer-wise manner [7–9]. Formally, we state the definition of scale invariance as follow.

Definition 9.1 (Scale invariance) Given the loss function $\ell(\mathbf{w}, \tilde{\theta})$ parameterized by $\mathbf{w}$ and $\tilde{\theta}$, we call that $\mathbf{w}$ is a **scale-invariant parameter** of ℓ if for all positive constant $\alpha > 0$, $\ell(\alpha \mathbf{w}, \tilde{\theta}) = \ell(\mathbf{w}, \tilde{\theta})$ [9]. We further say ℓ has **scale invariance** if it has non-null scale-invariant parameters.

Furthermore, a function with scale invariance has the following properties, which is important for stabilizing training.

Theorem 9.2 (scale invariance property) *Let $\mathbf{w}$ be a scale-invariant parameter of $\ell(\mathbf{w}, \tilde{\theta})$ and we use (stochastic) gradient based update:* $\mathbf{w}_{t+1} = \mathbf{w}_t - \eta_t \frac{\partial \ell}{\partial \mathbf{w}_t}$. *We have: (1)* $\frac{\partial \ell}{\partial \mathbf{w}} \cdot \mathbf{w} = 0$ *and* $\|\mathbf{w}_{t+1}\|_2^2 = \|\mathbf{w}_t\|_2^2 + \eta_t^2 \|\frac{\partial \ell}{\partial \mathbf{w}}\|_2^2$; *(2)* $\frac{\partial \ell}{\partial \mathbf{w}}|_{\mathbf{w}=\alpha \mathbf{w}_0} = \frac{1}{\alpha} \frac{\partial \ell}{\partial \mathbf{w}}|_{\mathbf{w}=\mathbf{w}_0}$, $\forall \alpha > 0$.

The scale invariance property ensures that gradient descent based methods always increase the norm of the weight [4, 5, 9–12], and furthermore the gradients w.r.t.weight get smaller as the norm of weight becomes larger. This stabilizes the training from diverging to infinity [9].

BN ensures the weight vector connecting to an output neuron to be a scale-invariant parameter. That is, $BN(\mathbf{x}; \alpha \mathbf{w}) = BN(\mathbf{x}; \mathbf{w})$ for $\alpha > 0$, where $\mathbf{w}$ is the weight vector that connects to the output neuron followed by BN. The scale-invariant property also applies to other methods that normalize the activations [3, 13–15] or weights [10, 16, 17]. This property was first shown in the original BN paper, and then investigated in [3] to compare different normalization methods, and further extended for rectifier networks in [18]. Here, we illustrate how BN stabilizes the training due to its scale-invariant property in a network. From the perspective of a practitioner, two phenomena relate to the instability in training a DNN: (1) the training loss first increases significantly and then diverges; or (2) the training loss hardly changes, compared to the initial condition. The former is mainly caused by weights with large updates (e.g., exploded gradients or optimization with a large learning rate). The latter is caused by weights with few updates (vanished gradients or optimization with a small learning rate). In the following theorem, we show that unnormalized rectifier neural network is very likely to encounter both phenomena.

Theorem 9.3 *Given a rectifier neural network with nonlinearity $\phi(\alpha \mathbf{x}) = \alpha \phi(\mathbf{x})$ ($\alpha > 0$), if the weight in each layer is scaled by $\widehat{W}_l = \alpha_l W_l$ ($l = 1, \ldots, L$ and $\alpha_l > 0$), we have* *the scaled layer input:* $\widehat{\mathbf{x}}_l = (\prod\limits_{i=1}^{l} \alpha_i)\mathbf{x}_l.$ *Assuming that* $\frac{\partial \mathcal{L}}{\partial \mathbf{h}_L} = \mu \frac{\partial \mathcal{L}}{\partial \mathbf{h}_L},$ *we have the output-*

gradient: $\frac{\partial \mathcal{L}}{\partial \widehat{\mathbf{h}}_l} = \mu(\prod\limits_{i=l+1}^{L} \alpha_i)\frac{\partial \mathcal{L}}{\partial \mathbf{h}_l},$ *and weight-gradient:* $\frac{\partial \mathcal{L}}{\partial \widehat{\mathbf{W}}_l} = (\mu \prod\limits_{i=1,i\neq l}^{L} \alpha_i)\frac{\partial \mathcal{L}}{\partial \mathbf{W}_l},$ *for all* $l =$
$1, \ldots, L.$

The proof is shown in Appendix A.3. From Theorem 9.3, we observe that the scaled factor α_l of the weight in layer l will affect all other layers' weight-gradients. Specifically, if all $\alpha_l > 1$ ($\alpha_l < 1$), the weight-gradient will increase (decrease) exponentially for one iteration. Moreover, such an exponentially increased weight-gradient will be sustained and amplified in the subsequent iteration, due to the increased magnitude of the weight caused by updating. That is why the unnormalized neural network will diverge, once the training loss increases over a few continuous iterations. We show that such instability can be relieved by BN, based on the following theorem.

Theorem 9.4 *Under the same condition as Theorem 9.3, for the normalized network with* $\mathbf{h}_l = \mathbf{W}_l\mathbf{x}_{l-1}$ *and* $\mathbf{s}_l = BN(\mathbf{h}_l),$ *we have:* $\widehat{\mathbf{x}}_l = \mathbf{x}_l,$ $\frac{\partial \mathcal{L}}{\partial \widehat{\mathbf{h}}_l} = \frac{1}{\alpha_l}\frac{\partial \mathcal{L}}{\partial \mathbf{h}_l},$ $\frac{\partial \mathcal{L}}{\partial \widehat{\mathbf{W}}_l} = \frac{1}{\alpha_l}\frac{\partial \mathcal{L}}{\partial \mathbf{W}_l},$ *for all* $l =$
$1, \ldots, L.$

The proof is shown in Appendix A.3. From Theorem 9.4, the scaled factor α_l of the weight will not affect other layers' activations/gradients. The magnitude of the weight-gradient is inversely proportional to the scaled factor. Such a mechanism will stabilize the weight growth/reduction, as shown in [19, 20].

Empirical Analysis. We conduct experiments to show how the activation/gradient is affected by initialization in unnormalized DNNs (indicated as 'Plain') and batch normalized DNNs (indicated as 'BN'). We train a 20-layer MLP, with 256 neurons in each layer, for MNIST classification. The nonlinearity is ReLU. We use the full gradient descent and report the results based on the best training loss among learning rates in {0.05, 0.1, 0.5, 1}. In Fig. 9.1a and b, we observe that the magnitude of the layer input (output-gradient) of 'Plain' for random initialization [21] suffers from exponential decrease during forward pass (backward pass). The main reason for this is that the weight has a small magnitude, based on Theorem 9.3. This problem can be relieved by He-initialization [22], where the magnitude of the input/output-gradient is stable across layers (Fig. 9.1c and d). We observe that BN can well preserve the magnitude of the input/output-gradient across different layers for both initialization methods.

Weight Domination. It was shown the scale-invariant property of BN has an implicit early stopping effect on the weight matrices [3], helping to stabilize learning towards convergence. Here, we show that this layer-wise 'early stopping' sometimes results in the false impression of a local minimum, which has detrimental effects on the learning, since the network does

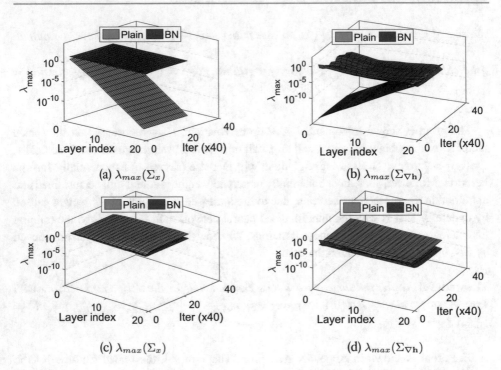

(a) $\lambda_{max}(\Sigma_x)$ (b) $\lambda_{max}(\Sigma_{\nabla \mathbf{h}})$

(c) $\lambda_{max}(\Sigma_x)$ (d) $\lambda_{max}(\Sigma_{\nabla \mathbf{h}})$

Fig. 9.1 Analysis of the magnitude of the layer input (indicated by $\lambda_{max}(\Sigma_x)$) and layer output-gradient (indicated by $\lambda_{max}(\Sigma_{\nabla \mathbf{h}})$). The experiments are performed on a 20-layer MLP with 256 neurons in each layer, for MNIST classification. The results of **a b** are under random initialization [21], while **c d** He-initialization [22]. Images with permission from [18]

not well learn the representation in the corresponding layer. For illustration, we provide a rough definition termed *weight domination*, with respect to a given layer.

Definition 9.5 Let $\mathbf{W}_l$ and $\frac{\partial \mathcal{L}}{\partial \mathbf{W}_l}$ be the weight matrix and its gradient in layer l. If $\lambda_{max}(\frac{\partial \mathcal{L}}{\partial \mathbf{W}_l}) \ll \lambda_{max}(\mathbf{W}_l)$, where $\lambda_{max}(\cdot)$ indicates the maximum singular value of a matrix, we refer to layer k has a state of **weight domination**.

Weight domination implies a smoother gradient with respect to the given layer. This is a desirable property for linear models (the distribution of the input is fixed), where the optimization objective targets to arrive the stationary points with smooth (zero) gradient. However, weight domination is not always desirable for a given layer of a DNN, since such a state of one layer is possibly caused by the increased magnitude of the weight matrix or decreased magnitude of the layer input (the non-convex optimization in Eq. 2.12), not necessary driven by the optimization objective itself. Although BN ensures a stable distribution of layer inputs, a network with BN still has the possibility that the magnitude

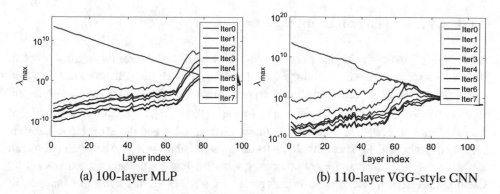

(a) 100-layer MLP   (b) 110-layer VGG-style CNN

Fig. 9.2 Experiments relating to gradient explosion of BN in deep networks without residual connections. We calculate the maximum eigenvalues of the sub-FIMs, and provide the results for the first seven iterations of (**a**) a 100-layer MLP for MNIST classification and (**b**) a 110-layer VGG-style CNN for CIFAR-10 classification. We observe that the weight-gradient has exponential explosion at initialization ('Iter0'). After a single step, the first-step gradients dominate the weights due to gradient explosion in lower layers, hence the exponential growth in the magnitude of the weight. This increased magnitude of weight leads to small weight gradients ('Iter1' to 'Iter7'), which is caused by BN, as discussed in the book. Therefore, some layers (especially the lower layers) of the network enter the state of *weight domination*. Images with permission from [18]

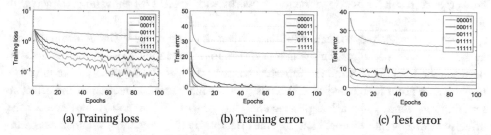

(a) Training loss   (b) Training error   (c) Test error

Fig. 9.3 Exploring the effectiveness of weight domination. We run the experiments on a 5-layer MLP with BN and the number of neuron in each layer is 256. We simulate weight domination in a given layer by blocking its weight updates. We denote '0' in the legend as the state of weight domination (the first digit represents the first layer). Images with permission from [18]

of the weight in a certain layer is significantly increased. We experimentally observe this phenomenon, as shown in Fig. 9.2.

Weight domination sometimes harms the learning of the network, because this state limits its ability to learn the representation in the corresponding layer. To investigate this, we conduct experiments on a 5-layer MLP and show the results in Fig. 9.3. We observe that the network with weight domination in certain layers, can still decrease the loss, but has degenerated performance.

9.1.1 Auto-Tuning on Learning Rate

Arora et al. [9] define the *intrinsic optimization problem* that optimizes the original problem with unit-norm constraint (combining Eqs. 7.2 and 7.3). It can be demonstrated that the *effective learning rate* for the scale-invariant weight parameter is $\frac{\eta_t}{\|\mathbf{w}_t\|_2^2}$, which has an auto-tuning effect corresponding to the scale-invariant parameter [9, 23–25]. Given the assumptions that the loss function $\ell(\mathbf{w}, \tilde{\theta})$ is twice continuously differentiable and the expected loss lower-bounded, Arora et al. [9] show that if the learning rate for $\tilde{\theta}$ is optimally set, then $(\mathbf{w}, \tilde{\theta})$ can converge to a first-order stationary point in (stochastic) gradient descent no matter what fix learning rate is set for the scale-invariant parameter $\mathbf{w}$. They further show that the rates of convergence are $O(T^{-1/2})$ and $O(T^{-1/4} polylog(T))$, for full-batch GD and SGD, respectively, where T is the iteration number. Cai et al. [24] analyze the simple problem of ordinary least squares (a linear model) with BN. They show that the iteration sequence $(\mathbf{w}, \tilde{\theta})$ using full batch GD with BN converges to a stationary point for any initial value and any learning rate $\eta_{\mathbf{w}} > 0$ for the scale-invariant parameter $\mathbf{w}$, as long as the learning rate $\eta_{\tilde{\theta}} \in (0, 1]$ for $\tilde{\theta}$. Besides, Kohler et al. [26] demonstrate that BN obtains an accelerated convergence on the (possibly) nonconvex problem of learning half-spaces with Gaussian inputs, from a length-direction decoupling perspective. Dukler et al. [27] further provide the first global convergence result for two-layer neural networks with ReLU [28] activations trained with weight normalization.

Based on the analyses of scale invariance, Wu et al. [11] propose WNgrad that introduces an adaptive step-size algorithm, which is shown to achieve robustness to the relationship between the learning rate and the Lipschitz constant. Heo et al. [29] observe that the additional momentum introduced in GD leads to a far more rapid reduction in the effective learning rate for scale-invariant parameters. They propose SGDP and AdamP to alter the effective learning rate without changing the effective update directions, which have advantages over the naive SGD and Adam with empirical validation on multiple benchmarks.

Another research direction is to analyze the effect of weight decay [30] when combined with normalization methods [6, 8, 31–35]. In this case, weight decay causes the parameters to have smaller norms, and thus the effective learning rate is larger. Li and Arora [35] show that the original learning rate schedule and weight decay can be folded into a new exponential schedule, when normalization methods with scale invariance property are used.

9.2 Improved Conditioning in Optimization

As stated in Chap. 2, one motivation behind BN is that whitening the input can improve the conditioning of the optimization [19] and thus accelerate training [14, 36]. This motivation is theoretically supported for linear models [35, 37] by connecting the curvature matrix (e.g., Hessian or FIM) to the covariance matrix of the input, and precisely characterizing the training dynamics using the spectrum of the curvature matrix. This analysis is further extended to DNNs by approximating the curvature matrix using the Kronecker product (K-FAC) [38], in which the full FIM can be represented as multiple independent sub-FIMs and each sub-FIM can be calculated as the Kronecker product between the covariance matrix of the layer input $\Sigma_{\mathbf{x}}^l$ and covariance matrix of layer output-gradient $\Sigma_{\nabla \mathbf{h}}^l$ (see Chap. 2 for details). It is clear that the conditioning of the FIM can be improved, if (1) the statistics of the layer input (e.g., $\Sigma_{\mathbf{x}}^l$) and output-gradient (e.g., $\Sigma_{\nabla \mathbf{h}}^l$) across different layers are equal; or (2) $\Sigma_{\mathbf{x}}^l$ and $\Sigma_{\nabla \mathbf{h}}^l$ are well conditioned.

Daneshmand et al. [39] show that BN prevents the rank collapse of the covariance matrix of the layer input $\Sigma_{\mathbf{x}}^l$. In particular, they theoretically demonstrate that the covariance matrix of the deepest layer $\Sigma_{\mathbf{x}}^L$ has a rank at least as large as $\Omega(\sqrt{d})$ (d is the width of the network) in a deep linear neural network with BN, given the assumptions that: (1) the input needs to be a full rank; (2) the weight matrices are randomly initialized from a zero-mean, unit-variance distribution whose law is symmetric around zero. When given the same assumptions, the sequences of covariance matrix of layer input $\{\Sigma_{\mathbf{x}}^l\}$ converges to a rank one matrix in an unnormalized linear networks. Lubana et al. [40] further extend this claim in a deep linear neural network with group normalization (GN), showing that $\Sigma_{\mathbf{x}}^L$ has a rank at least as large as $\Omega(\sqrt{d/g})$, where g denotes the group size of GN. Note that the stable-rank [39] of a matrix can be used to implicitly characterize the conditioning of a matrix. The collapsed representation will lose the information during forward propagation, and make different inputs indistinguishable, which significantly obstructs the optimization [40]. These claims are also empirically validated by investigating the correlations between expressivity of activations and training dynamics [40], either using BN or other alternative normalization methods (e.g., GN, LN and IN).

Santurkar et al. [41] argue that BN may improve optimization by enhancing the smoothness of the Hessian of the loss. However, this conclusion is based on a layer-wise analysis [18, 41], which corresponds to the diagonal blocks of the overall Hessian. It is disputed by the empirical studies in [42], showing the exact opposite behavior of BN on a ResNet-20 network. Ghorbani et al. [43] further empirically investigate the conditioning of the optimization problem by computing the spectrum of the Hessian for a large-scale dataset. It is believed that the improved conditioning enables large learning rates for training, thus improving the generalization, as shown in [44]. Karakida et al. [45] investigate the conditioning of the optimization problem by analyzing the geometry of the parameter space determined by the FIM, which also corresponds to the local shape of the loss landscape under certain conditions.

One intriguing phenomenon is that the theoretical benefits of whitening the input for optimization only hold when BN is placed before the linear layer, while, in practice, BN is typically placed after the linear layer, as recommended in [19]. Huang et al. [18] experimentally observed, through a layer-wise conditioning analysis, that BN (placed after the linear layer) not only improves the conditioning of the activation's covariance matrix, but also improves the conditioning of the output-gradient's covariation. Similar observations were made in [39], where BN prevents the rank collapse of pre-activation matrices. Some works have also empirically investigated the position, at which BN should be plugged in

[14, 46, 47]. Results have shown that placing it after the linear layer may work better, in certain situations.

Other analyses of normalization in optimization include an investigation into the signal propagation and gradient backpropagation [20, 48, 49], based on the mean field theory [20, 48, 50]. Besides, the work of [26] demonstrated that BN obtains an accelerated convergence on the (possibly) nonconvex problem of learning half-spaces with Gaussian inputs, from a length-direction decoupling perspective. Dukler et al. [27] further provided the first global convergence result for two-layer neural networks with ReLU [28] activations trained with weight normalization.

9.3 Stochasticity for Generalization

One important property of BN is its ability to improve the generalization of DNNs. It is believed such an improvement is obtained from the stochasticity/noise introduced by normalization over batch data [19, 51, 52].

It is clear that both the normalized output (Eq. 3.5) and the population statistics (Eq. 3.6) can be viewed as stochastic variables, because they depend on the mini-batch inputs, which are sampled over datasets (Fig. 9.4). Therefore, the stochasticity comes from the normalized output during training [53], and the discrepancy of normalization between training (using estimated population statistics) and inference (using estimated population statistics) [54,

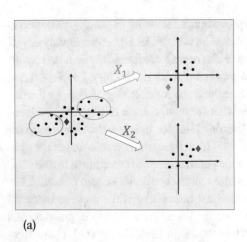
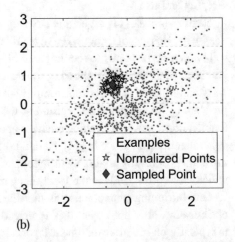

(a)   (b)

Fig. 9.4 Illustration of the stochasticity introduced by normalization over mini-batch data. (**a**) The representation of an example vary, when combined with different mini-batch for normalization. (**b**) We sample 1000 examples (black points) from a Gaussian distribution in a 16-dimensional space, and show the examples in the two-dimension sub-space (the 6th and 16th dimension). Given an example $\mathbf{x}$ (red diamond), when combining with 100 different mini-batches $\mathbf{X}^B$ ($B = 64$), we provide the normalized output $\hat{\mathbf{x}}$ (yellow pentagram) for BN

55]. Ioffe and Szegedy [19] were the first to show the advantages of this stochasticity for the generalization of networks, like dropout [56, 57]. Teye et al. [58] demonstrated that training a DNN using BN is equivalent to approximating inference in Bayesian models, and that uncertainty estimates can be obtained from any network using BN through Monte Carlo sampling during inference. This idea was further efficiently approximated by stochastic batch normalization [59] and exploited in prediction-time batch normalization [60]. We will illustrate the works addressing to model the stochasticity either by theoretic or empirical analyses.

9.3.1   Theoretical Model for Stochasticity

Shekhovtsov and Flach [51] jointly formulate the stochasticity of the normalized output and the discrepancy of normalization between training and inference in a mathematical way, under the assumptions that the distribution of activations over the full dataset is approximately Gaussian and $i.i.d.$ Specifically, given the assumptions: (1) the distribution of network activations X over the full dataset is approximately normal with statistics $(\hat{\mu}, \hat{\sigma}^2)$; (2) the activations for different training inputs are $i.i.d$, the train-time BN can be written as:

$$\frac{x - \mu_{\mathcal{B}}}{\sigma_{\mathcal{B}}} = (\frac{x - \hat{\mu}}{\hat{\sigma}} + \frac{\hat{\mu} - \mu_{\mathcal{B}}}{\hat{\sigma}})\frac{\hat{\sigma}}{\sigma_{\mathcal{B}}}. \tag{9.1}$$

Using the above assumptions, we have $\mu_{\mathcal{B}}$ being a random variable distributed as $\mathcal{N}(\hat{\mu}, \frac{1}{m}\hat{\sigma}^2)$. It can be demonstrated that:

$$\frac{\hat{\mu} - \mu_{\mathcal{B}}}{\hat{\sigma}} \sim \frac{1}{\sqrt{m}}\mathcal{N}(0, 1), \ \frac{\sigma^2}{\hat{\sigma}^2} \sim \frac{1}{m}\chi^2_{m-1} \ and \ \frac{\hat{\sigma}}{\sigma} \sim \sqrt{m}\chi^{-1}_{m-1}, \tag{9.2}$$

where χ^2 is chi-square distribution and χ^{-1} is the inverse chi distribution. Therefore, the train-time BN can be viewed as its test-time normalization ($\frac{x - \hat{\mu}}{\hat{\sigma}}$) combining with two random variances ($\frac{\hat{\mu} - \mu_{\mathcal{B}}}{\hat{\sigma}} \sim \frac{1}{\sqrt{m}}\mathcal{N}(0, 1)$ and $\frac{\hat{\sigma}}{\sigma} \sim \sqrt{m}\chi^{-1}_{m-1}$). Shekhovtsov and Flach [51] experimentally observe that the real statistics $\frac{\hat{\mu} - \mu_{\mathcal{B}}}{\hat{\sigma}}$ and $\frac{\hat{\sigma}}{\sigma}$, calculated from randomly drawn mini-batch samples, are close to the theoretical prediction.

Stochastic Axis Swapping As discussed in Sect. 4.2.1, PCA whitening over batch data suffers significant instability in training DNNs, and hardly converges, due to the so called stochastic axis swapping (SAS). Here, we provide a illustration as Given a data point represented as a vector $\mathbf{x} \in \mathbb{R}^d$ under the standard basis, its representation under another orthogonal basis $\{\mathbf{d}_1, \ldots, \mathbf{d}_d\}$ is $\hat{\mathbf{x}} = \mathbf{D}^T\mathbf{x}$, where $\mathbf{D} = [\mathbf{d}_1, \ldots, \mathbf{d}_d]$ is an orthogonal matrix. We define *stochastic axis swapping* as follows:

Definition 9.6 Assume a training algorithm that iteratively update weights using a batch of randomly sampled data points per iteration. **Stochastic axis swapping** occurs when a data point $\mathbf{x}$ is transformed to be $\hat{\mathbf{x}}_1 = \boldsymbol{D}_1^T \mathbf{x}$ in one iteration and $\hat{\mathbf{x}}_2 = \boldsymbol{D}_2^T \mathbf{x}$ in another iteration such that $\boldsymbol{D}_1 = \boldsymbol{P} \boldsymbol{D}_2$ where $\boldsymbol{P} \neq \boldsymbol{I}$ is a permutation matrix solely determined by the statistics of a batch.

Stochastic axis swapping makes training difficult, because the random permutation of the input dimensions can greatly confuse the learning algorithm—in the extreme case where the permutation is completely random, what remains is only a bag of activation values (similar to scrambling all pixels in an image), potentially resulting in an extreme loss of information and discriminative power.

Here, we demonstrate that the whitening of activations, if not done properly, can cause *stochastic axis swapping* in training neural networks. We start with standard PCA whitening [36], which computes $\Sigma^{-1/2}$ through eigen decomposition: $\Sigma_{pca}^{-1/2} = \widetilde{\Lambda}^{-1/2} \boldsymbol{D}^T$, where $\widetilde{\Lambda} = \mathrm{diag}(\lambda_1, \ldots, \lambda_d)$ and $\boldsymbol{D} = [\mathbf{d}_1, \ldots, \mathbf{d}_d]$ are the eigenvalues and eigenvectors of Σ, i.e. $\Sigma = \boldsymbol{D} \widetilde{\Lambda} \boldsymbol{D}^T$. That is, the original data point (after centering) is rotated by $\boldsymbol{D}^T$ and then scaled by $\widetilde{\Lambda}^{-1/2}$. Without loss of generalization, we assume that $\mathbf{d}_i$ is unique by fixing the sign of its first element. A first opportunity for stochastic axis swapping is that the columns (or rows) of $\widetilde{\Lambda}$ and $\boldsymbol{D}$ can be permuted while still giving a valid whitening transformation. But this is easy to fix—we can commit to a unique $\widetilde{\Lambda}$ and $\boldsymbol{D}$ by ordering the eigenvalues non-increasingly.

But it turns out that ensuring a unique $\widetilde{\Lambda}$ and $\boldsymbol{D}$ is insufficient to avoid stochastic axis swapping. Figure 9.5 illustrates an example. Given a mini-batch of data points in one iteration as shown in Fig. 9.5a, PCA whitening rotates them by $\boldsymbol{D}^T = [\mathbf{d}_1^T, \mathbf{d}_2^T]^T$ and stretches them along the new axis system by $\widetilde{\Lambda}^{-1/2} = diag(1/\sqrt{\lambda_1}, 1/\sqrt{\lambda_2})$, where $\lambda_1 > \lambda_2$. Considering another iteration shown in Fig. 9.5b, where all data points except the red points are the same,

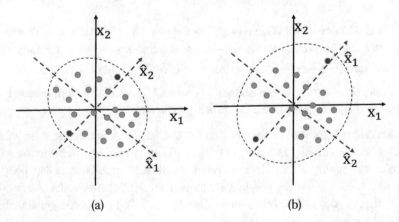

(a)   (b)

Fig. 9.5 Illustration that PCA whitening suffers from stochastic axis swapping. (**a**) The axis alignment of PCA whitening in the initial iteration; (**b**) The axis alignment in another iteration. Images with permission from [14]

it has the same eigenvectors with different eigenvalues, where $\lambda_1 < \lambda_2$. In this case, the new rotation matrix is $(D')^T = [d_2^T, d_1^T]^T$ because we always order the eigenvalues non-increasingly. The blue data points thus have two different representations with the axes swapped.

9.3.2 Empirical Analyses for Stochasticity

Even though the theoretical model for stochasticity is appealing, the required assumptions are usually not satisfied in practice. Huang et al. [53] proposed an empirical evaluation for the stochasticity of normalization over batch data, called stochastic normalization disturbance (SND), and investigated how the batch size affects the stochasticity of BN.

Stochastic Normalization Disturbance Given a sample $\mathbf{x} \in \mathbb{R}^d$ from a distribution P_χ, we take a sample set $\mathbf{X}^B = \{\mathbf{x}_1, \ldots, \mathbf{x}_B, \mathbf{x}_i \sim P_\chi\}$ with a size of B. We denote the normalization operation as $F(\cdot)$ and the normalized output as $\hat{\mathbf{x}} = F(\mathbf{X}^B; \mathbf{x})$. For a certain $\mathbf{x}$, $\mathbf{X}^B$ can be viewed as a random variable [58, 59]. $\hat{\mathbf{x}}$ is thus a random variable which shows the stochasticity. It's interesting to explore the statistical momentum of $\mathbf{x}$ to measure the magnitude of the stochasticity. Here we define the *Stochastic Normalization Disturbance* (SND) for the sample $\mathbf{x}$ over the normalization $F(\cdot)$ as:

$$\mathbf{\Delta}_F(\mathbf{x}) = \mathbb{E}_{\mathbf{X}^B}(\|\hat{\mathbf{x}} - \mathbb{E}_{\mathbf{X}^B}(\hat{\mathbf{x}})\|_2). \tag{9.3}$$

It's difficult to accurately compute this momentum if no further assumptions are made over the random variable $\mathbf{X}^B$, however, we can explore its empirical estimation over the sampled sets as follows:

$$\widehat{\mathbf{\Delta}}_F(\mathbf{x}) = \frac{1}{s} \sum_{i=1}^{s} \| F(\mathbf{X}_i^B; \mathbf{x}) - \frac{1}{s} \sum_{j=1}^{s} F(\mathbf{X}_j^B; \mathbf{x}) \|, \tag{9.4}$$

where s denotes the time of sampling. Figure 9.6 gives the illustration of sample $\mathbf{x}$'s SND with respect to the operation of BN. We can find that SND is closely related to the batch size. When batch size is large, the given sample $\mathbf{x}$ has a small value of SND and the transformed outputs have a compact distribution. As a consequence, the stochastic uncertainty $\mathbf{x}$ can be low.

SND can be used to evaluate the stochasticity of a sample after the normalization operation, which works like the dropout rate [56]. We can further define the normalization operation $F(\cdot)$'s SND as: $\mathbf{\Delta}_F = \mathbb{E}_{\mathbf{x}}(\mathbf{\Delta}(\mathbf{x}))$ and it's empirical estimation as $\widehat{\mathbf{\Delta}}_F = \frac{1}{N} \sum_{i=1}^{N} \widehat{\mathbf{\Delta}}(\mathbf{x})$ where N is the number of sampled examples. $\mathbf{\Delta}_F$ describes the magnitudes of stochasticity for corresponding normalization operations.

Exploring the exact statistic behavior of SND is difficult and out of the scope of this book. We can, however, explore the relationship of SND related to the batch size and feature dimension. We find that our defined SND gives a reasonable explanation to why we should control the extent of whitening and why mini-batch based normalizations have a degenerate performance when given a small batch size.

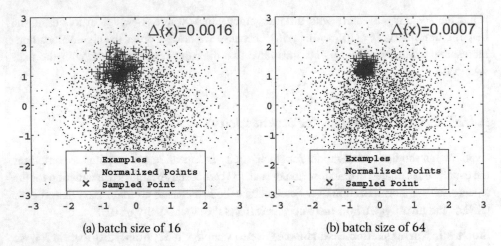

(a) batch size of 16   (b) batch size of 64

Fig. 9.6 Illustration of SND with different batch sizes. We sample 3000 examples (black points) from Gaussian distribution. We show a given example $\mathbf{x}$ (red cross) and its BN outputs (blue plus sign), when normalized over different sample sets $\mathbf{X}^B$. (**a**) and (**b**) show the results with batch sizes B of 16 and 64, respectively. Images with permission from [53]

This empirical analysis was further extended to the more general BW in [55]. Here, we conduct experiments to quantitatively evaluate the effects of different normalization methods. Noticeably, the stochasticity is related to the batch size m and the dimension d. Figure 9.7a shows the SND of different normalization methods with respect to the dimensions, when fixing the batch size to 1024. We find that PCA whitening shows the largest SND while BN the smallest, over all the dimensions. We notice that all whitening methods have an increased SND when the dimension increases. Besides, ZCA has a smaller SND than CD, over all the dimensions. Figure 9.7b shows the SND of different normalization methods with respect to the batch size, when fixing the dimension to 128. An interesting observation is that PCA whitening has nearly the same large SND among different batch sizes. This suggests that the PCA whitening is extremely unstable, no matter how accurate the estimation of the mini-batch covariance matrix is. This effect is in accordance with the explanation of Stochastic Axis Swapping (SAS) shown in [14], where a small change over the examples (when performing PCA whitening) results in a large change of representation.

To further investigate how this stochasticity affects DNN training, we perform experiments on a four-layer Multilayer Perceptron (MLP) with 256 neurons in each layer. We evaluate the training loss with respect to the epochs, and show the results in Fig. 9.8a. We find that, among all the whitening methods, ZCA works the best, while PCA is the worst. We argue that this correlates closely with the SND they produce. Apparently, the increased stochasticity can slow down training, even though all the whitening methods have equivalently improved conditioning. An interesting observation is that, in this case, BN works better than ZCA whitening. This is surprising since ZCA has improved conditioning over

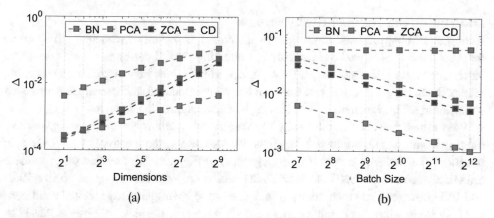

Fig. 9.7 SND comparison of different batch whitening methods. We sample 60,000 examples from a Gaussian distribution as the training set. To calculate SND, we use $s = 200$ and $N = 20$. We show (**a**) the SND with respect to the dimensions ranging from 2^1 to 2^9, under a batch size of 1024; (**b**) the SND with respect to the batch size ranging from 2^7 to 2^{12}, under a dimension of 128. Images with permission from [55]

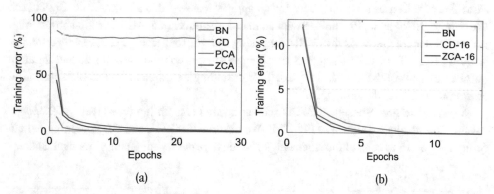

Fig. 9.8 Experiments on a 4-layer MLP with 256 neurons in each layer, for MNIST classification. We use a batch size of 1024 and report the training errors. (**a**) The results of full whitening methods; (**b**) The results of group based whitening, where 'ZCA-16' indicates ZCA whitening with a group size of 16. Images with permission from [55]

BN by removing the correlation [14], and it should theoretically have a better optimization behavior. However, the amplified stochasticity of ZCA whitening mitigates this advantage in optimization, thus resulting in a degenerated performance. Therefore, from an optimization perspective, we should control the extent of the stochasticity.

Controlling the Stochasticity by Groups

Huang et al. [14] proposed to use groups to control the extent of whitening. They argue that this method reduces the inaccuracy in estimating the full covariance matrix when the batch size is not sufficiently large. Here, we empirically show how group based whitening affects the SND, providing a good trade-off between introduced stochasticity and improved conditioning. This is essential for achieving a better optimization behavior.

We evaluate the SND of different whitening transformations by varying the group size ranging from 2 to 512, as shown in Fig. 9.9a. We also display the spectrum of the covariance matrix of the whitened output (based on groups) in Fig. 9.9b. We find that the group size effectively controls the SND of the ZCA/CD whitening. With decreasing group size, ZCA and CD show reduced stochasticity (Fig. 9.9a), while having also degenerated conditioning (Fig. 9.9b), since the output is only partially whitened. Besides, we observe that PCA whitening still has a large SND over all group sizes, and with no significant differences. This observation further corroborates the explanation of SAS given in [14], i.e., that the PCA whitening is extremely unstable.

We also show the SND of the approximate ZCA whitening method (called ItN [53]) in Fig. 9.9a, which uses Newton's iteration to approximately calculate the whitening matrix. We denote 'ItN5' as the ItN method with an iteration number of 5. An interesting observation is that ItN has smaller SND than BN, when using a large group size (e.g., 256) with a smaller iteration (e.g., T = 5). This suggests that we can further combine group size and iteration number to control the stochasticity for ItN, providing an efficient and stable solution to approximate ZCA whitening [53].

We also use group-based ZCA/CD whitening methods on the four-layer MLP experiments. The results are shown in Fig. 9.8b. We observe that ZCA and CD whitening, with a group size of 16 to control the stochasticity, achieve better training behaviors than BN.

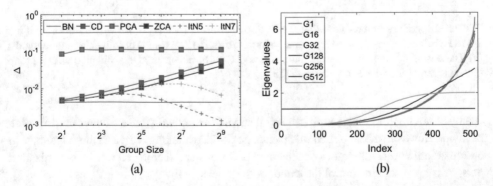

(a)   (b)

Fig. 9.9 Group-based whitening experiments. (**a**) We show the SND of different normalization operations with respect to the group size. The experimental setup is the same as Fig. 9.7 and the input dimension is $d = 512$. (**b**) We show the spectrum of covariance matrix of the ZCA whitened output (Note that CD/PCA whitening has the same spectrum as ZCA whitening.), where 'G16' indicates whitening with a group size of 16. Images with permission from [55]

Table 9.1 Summary of $\zeta(\phi; \mathbf{X})$, $\zeta(\phi; \mathbb{D})$ and ranges of m/g for normalization methods. The analysis can be naturally extended to CNN, following how BN (GN) extents from MLP to CNN shown in Chap. 4. For examples, the number of neurons to be normalized for GN/GW is $d = d'HW$, and the number of samples to be normalized for BN/BW is $m = m'HW$, given the input $\mathbf{X} \in \mathbb{R}^{d' \times m' \times H \times W}$ for CNN. The results with permission from [64]

	Along a batch		Along a group of neurons	
	BN	BW	GN	GW
$\zeta(\phi; \mathbf{X})$	$2d$	$\frac{d(d+3)}{2}$	$2gm$	$\frac{mg(g+3)}{2}$
$\zeta(\phi; \mathbb{D})$	$\frac{2Nd}{m}$	$\frac{Nd(d+3)}{2m}$	$2gN$	$\frac{Ng(g+3)}{2}$
Ranges of m/g	$m \geq 2$	$m \geq \frac{d+3}{2}$	$g \leq \frac{d}{2}$	$g \leq \frac{\sqrt{8d+9}-3}{2}$

Some studies exploit the stochasticity of BN to improve the generalization for large-batch training, by altering the batch size when estimating the population statistics. One typical work is ghost batch normalization [61–63], which reduces the generalization error by acquiring the statistics on small virtual ('ghost') batches instead of the real large batch.

9.4 Effects on Representation

The normalization operation ensures that the normalized output $\widehat{\mathbf{X}} = \phi(\mathbf{X}) \in \mathbb{R}^{d \times m}$ has a stable distribution. This stability of distribution can be implicitly viewed as the constraints imposed on $\widehat{\mathbf{X}}$, which can be represented as a system of equations $\Upsilon_\phi(\widehat{\mathbf{X}})$. For example, BN provides the constraints $\Upsilon_{\phi_{BN}}(\widehat{\mathbf{X}})$ as:

$$\sum_{j=1}^{m} \widehat{\mathbf{X}}_{ij} = 0 \text{ and } \sum_{j=1}^{m} \widehat{\mathbf{X}}_{ij}^2 - m = 0, \text{ for } i = 1, \ldots, d. \qquad (9.5)$$

We define the **constraint number** of normalization to quantitatively measure the magnitude of the constraints provided by the normalization method.

Definition 9.7 Given the input data $\mathbf{X} \in \mathbb{R}^{d \times m}$, the **constraint number** of a normalization operation $\phi(\cdot)$, referred to as $\zeta(\phi; \mathbf{X})$, is the number of independent equations in $\Upsilon_\phi(\widehat{\mathbf{X}})$. As an example, we have $\zeta(\phi_{BN}; \mathbf{X}) = 2d$ from Eq. 9.5. Furthermore, given training data $\mathbb{D}$ of size N, we consider the optimization algorithm with batch size m (we assume N is divisible by m). We calculate the constraint number of normalization over the entire training data $\zeta(\phi; \mathbb{D})$. Table 9.1 summarizes the constraint numbers of certain normalization methods discussed in this book (please refer to the Appendix A.2 for derivation details).

We can see that the whitening operation provides significantly stronger constraints than the standardization operation. Besides, the constraints get stronger for BN (GN), when reducing (increasing) the batch size (group number).

9.4.1   Constraint on Feature Representation

BN's benefits in accelerating the training of DNNs are mainly attributed to two reasons: (1) The distribution is more stable when fixing the first and second momentum of the activations, which reduces the internal covariant shifts [19]; (2) The landscape of the optimization objective is better conditioned [14, 41], by improving the conditioning of the activation matrix with normalization. Based on these arguments, GW/GN should have better performance when increasing the group number, due to the stronger constraints and better conditioning. However, Huang et al. [64] experimentally observe that GN/GW has significantly degenerated performance when the group number is too large, which is similar to the small-batch-size problem of BN/BW. For understanding the reason behind this phenomenon, we first show that the batch size/group number has a value range, which can be mathematically derived. The normalization operation can be regarded as a way to find a solution $\widehat{\mathbf{X}}$ satisfying the constraints $\Upsilon_\phi(\widehat{\mathbf{X}})$. To ensure the solution is feasible, it must satisfy the following condition:

$$\zeta(\phi; \mathbf{X}) \le \chi(\widehat{\mathbf{X}}), \tag{9.6}$$

where $\chi(\widehat{\mathbf{X}}) = md$ is the number of variables in $\widehat{\mathbf{X}}$. Based on Eq. 9.6, we have $m >= 2$ for BN to ensure a feasible solution. We also provide the ranges of batch size/group number for other normalization methods in Table 9.1. Note that the batch size m should be larger than or equal to d to achieve a numerically stable solution for BW when using ZCA whitening in practice [14]. This also applies to GW, where g should be less than or equal to $\sqrt{d}$.

We then demonstrate that normalization eventually affects the feature representation in a certain layer. Figure 9.10 shows the histogram of normalized output $\widehat{\mathbf{X}}$, by varying the channel number in each group c of GN[1] and batch size m of BN. We observe that: (1) the values of $\widehat{\mathbf{X}}$ are heavily constrained if c or m is too small, e.g., the value of $\widehat{\mathbf{X}}$ is constrained to be $\{-1, +1\}$ if $c = 2$; (2) $\widehat{\mathbf{X}}$ is not Gaussian if c or m is too small, while BN/GN aims to produce a normalized output with a Gaussian distribution. We believe that the constrained feature representation caused by GN/GW with a large group number is the main factor leading to the degenerated performance of a network. Besides, we also observe that the normalized output of GN is more correlated than that of BN, which supports the claim that BN is more capable of improving the conditioning of activations than GN.

---

[1] Note that the channel number in each group $c = \frac{d}{g}$. We vary c, rather than g, for simplifying the discussion.

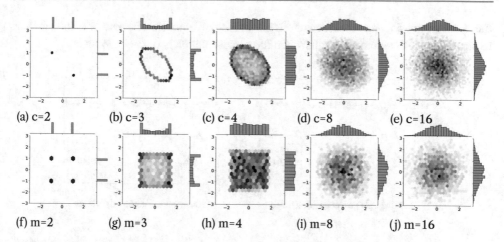

Fig. 9.10 Illustration of the normalized output of GN/BN. We perform normalization over 1,680 examples sampled from a Gaussian distribution, varying the channel number for each group c of GN (the upper subfigures) and the batch size m of BN (the lower subfigures). We plot the bivariate histogram (using hexagonal bins) of the normalized output in the two-dimensional subspace, and marginal histograms (using rectangular bins) in the one-dimensional subspace. Images with permission from [64]

We also seek to quantitatively measure the representation of a feature space. Given a set of features $\widetilde{\mathbb{D}} \in \mathbb{R}^{d \times N}$ extracted by a network, we assume the examples of $\widetilde{\mathbb{D}}$ belong in a d-dimensional hypercube $V = [-1, 1]^d$ (we can ensure that this assumption holds by dividing the maximum absolute value of each dimension). Intuitively, a powerful feature representation implies that the examples from $\widetilde{\mathbb{D}}$ spread over V with large diversity, while a weak representation indicates that they are limited to certain values without diversity. We thus define the diversity of $\widetilde{\mathbb{D}}$ based on the information entropy as follows, which can empirically indicate the representation ability of the feature space to some degree:

$$\Gamma_{d,T}(\widetilde{\mathbb{D}}) = \sum_{i=1}^{T^d} p_i \log p_i. \tag{9.7}$$

Here, V is evenly divided into T^d bins, and p_i denotes the probability of an example belonging to the i-th bin. We can thus calculate $\Gamma_{d,T}(\widetilde{\mathbb{D}})$ by sampling enough examples. However, calculating $\Gamma_{d,T}(\widetilde{\mathbb{D}})$ with reasonable accuracy requires $O(T^d)$ examples to be sampled from a d-dimensional space. We thus only calculate $\Gamma_{2,T}(\widetilde{\mathbb{D}})$ in practice by sampling two dimensions, and average the results. We show the diversity of group (batch) normalized features by varying the channels of each group (batch size) in Fig. 9.11, from which we can obtain similar conclusions as in Fig. 9.10.

In summary, our qualitative and quantitative analyses show that group/batch based normalizations have low diversity of feature representations when c/m is small. We believe these constrained feature representations affect the performance of a network, and can lead to significantly deteriorated results when the representation is over-constrained.

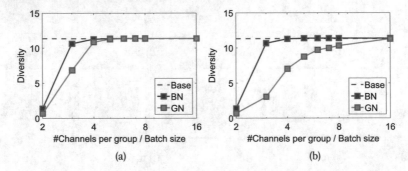

Fig. 9.11 Diversity of group (batch) normalized features when varying the channels per group (batch size). We sample $N = 1,680,000$ examples and use $1,000^2$ bins. We use the sampled Gaussian data as features in (**a**) and the output of a one-layer MLP in (**b**). Here, 'Base' indicates the diversity of unnormalized features. Images with permission from [64]

9.4.2   Effect on Representational Capacity of Model

The constraints introduced by normalization are believed to affect the representational capacity of neural networks [19], and thus the learnable scale and shift parameters are used to recover the representations [3, 13, 14, 19]. However, such an argument is seldom validated by either theoretical or empirical analysis. Theoretically analyzing the complexity measure (e.g., VC dimensions [66] or the number of linear regions [1, 2]) of the representational capacity of neural networks with normalization is a challenging task, because normalized networks do not follow the assumptions for calculating linear regions or VC dimensions. Here, we conduct preliminary experiments, seeking to empirically show how normalization affects the representational capacity of a network, by varying the constraints imposed on the feature.

We follow the non-parametric randomization tests fitting random labels [65] to empirically compare the representational capacity of neural networks. To rule out the optimization benefits introduced by normalization, we first conduct experiments using a linear classifier, where normalization is also inserted after the linear module. We train over 1,000 epochs using stochastic gradient descent (SGD) with a batch size of 16, and report the best training accuracy among the learning rates in $\{0.001, 0.005, 0.01, 0.05, 0.1\}$ in Fig. 9.12a. We observe that GN and GW have lower training accuracy than when normalization is not used, which suggests that normalization does indeed reduce the model's representational capacity in this case. Besides, the accuracy of GN/GW decreases as the group number increases. This suggests that the model may have weaker representational ability when increasing the constraints on the feature. Note that we have the same observations regardless of whether or not the learnable scale and shift parameters of GN/GW are used.

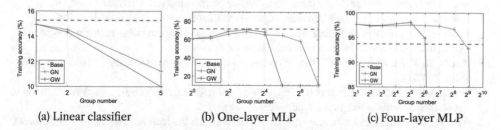

| (a) Linear classifier | (b) One-layer MLP | (c) Four-layer MLP |

Fig. 9.12 Comparison of model representational capacity when fitting random labels [65] on MNIST dataset using different architectures. We vary the group number of GN/GW and evaluate the training accuracy. 'Base' indicates the model without normalization. (**a**) Linear classifier; (**b**) One-layer MLP with 256 neurons in each layer; (**c**) Four-layer MLP with 1,280 neurons in each layer. Images with permission from [64]

To further consider the trade-off between the benefits of normalization on optimization and its constraints on representation, we conduct experiments on the one-layer and four-layer MLPs. The results are shown in Fig. 9.12b and c, respectively. We observe that the model with GN/GW has significantly degenerated training accuracy when g is too large, which means that a large group number heavily limits the model's representational capacity by constraining the feature representation, as discussed in Sect. 9.4.2. We note that GW is more sensitive to the group number than GN. The main reason is that $\zeta(\phi_{GW}; \mathbb{D})$ is quadratic to g, while $\zeta(\phi_{GN}; \mathbb{D})$ is linear to it, from Table 9.1. Besides, we observe that GN and GW still have lower training accuracy than 'Base' on the one-layer MLP, but higher accuracy on the four-layer MLP if the group number g is not too large. This suggests that the benefits of normalization on optimization dominate if the model's representation is not too limited. We also observe that the best training accuracy of GW is higher than that of GN. We attribute this to the fact that the whitening operation is better for improving the conditioning of optimization, compared to standardization.

References

1. Montufar, G. F., R. Pascanu, K. Cho, and Y. Bengio (2014). On the number of linear regions of deep neural networks. In *NeurIPS*.
2. Xiong, H., L. Huang, M. Yu, L. Liu, F. Zhu, and L. Shao (2020). On the number of linear regions of convolutional neural networks. In *ICML*.
3. Ba, L. J., R. Kiros, and G. E. Hinton (2016). Layer normalization. arXiv preprint arXiv:1607.06450.
4. Neyshabur, B., R. Tomioka, R. Salakhutdinov, and N. Srebro (2016). Data-dependent path normalization in neural networks. In *ICLR*.

5. Sun, J., X. Cao, H. Liang, W. Huang, Z. Chen, and Z. Li (2020). New interpretations of normalization methods in deep learning. In *AAAI*.

6. Wan, R., Z. Zhu, X. Zhang, and J. Sun (2020). Spherical motion dynamics of deep neural networks with batch normalization and weight decay. arXiv preprint arXiv:2006.08419.

7. Cho, M. and J. Lee (2017). Riemannian approach to batch normalization. In *NeurIPS*.

8. Hoffer, E., R. Banner, I. Golan, and D. Soudry (2018). Norm matters: efficient and accurate normalization schemes in deep networks. In *NeurIPS*.

9. Arora, S., Z. Li, and K. Lyu (2019). Theoretical analysis of auto rate-tuning by batch normalization. In *ICLR*.

10. Salimans, T. and D. P. Kingma (2016). Weight normalization: A simple reparameterization to accelerate training of deep neural networks. In *NeurIPS*.

11. Wu, X., R. Ward, and L. Bottou (2018). Wngrad: Learn the learning rate in gradient descent. arXiv preprint arXiv:1803.02865.

12. Roburin, S., Y. de Mont-Marin, A. Bursuc, R. Marlet, P. Pérez, and M. Aubry (2020). Spherical perspective on learning with batch norm. arXiv preprint arXiv:2006.13382.

13. Wu, Y. and K. He (2018). Group normalization. In *ECCV*.

14. Huang, L., D. Yang, B. Lang, and J. Deng (2018). Decorrelated batch normalization. In *CVPR*.

15. Ulyanov, D., A. Vedaldi, and V. S. Lempitsky (2016). Instance normalization: The missing ingredient for fast stylization. arXiv preprint arXiv:1607.08022.

16. Huang, L., X. Liu, Y. Liu, B. Lang, and D. Tao (2017). Centered weight normalization in accelerating training of deep neural networks. In *ICCV*.

17. Huang, L., X. Liu, B. Lang, A. W. Yu, Y. Wang, and B. Li (2018). Orthogonal weight normalization: Solution to optimization over multiple dependent stiefel manifolds in deep neural networks. In *AAAI*.

18. Huang, L., J. Qin, L. Liu, F. Zhu, and L. Shao (2020). Layer-wise conditioning analysis in exploring the learning dynamics of DNNs. In *ECCV*.

19. Ioffe, S. and C. Szegedy (2015). Batch normalization: Accelerating deep network training by reducing internal covariate shift. In *ICML*.

20. Yang, G., J. Pennington, V. Rao, J. Sohl-Dickstein, and S. S. Schoenholz (2019). A mean field theory of batch normalization. In *ICLR*.

21. LeCun, Y., L. Bottou, G. B. Orr, and K.-R. Muller (1998). Efficient backprop. In *Neural Networks: Tricks of the Trade*.

22. He, K., X. Zhang, S. Ren, and J. Sun (2015). Delving deep into rectifiers: Surpassing human-level performance on imagenet classification. In *ICCV*.

23. Wu, S., G. Li, L. Deng, L. Liu, Y. Xie, and L. Shi (2018). L1-norm batch normalization for efficient training of deep neural networks. arXiv preprint arXiv:1802.09769.

24. Cai, Y., Q. Li, and Z. Shen (2019). A quantitative analysis of the effect of batch normalization on gradient descent. In *ICML*, pp. 882–890.

25. Chai, E., M. Pilanci, and B. Murmann (2020). Separating the effects of batch normalization on CNN training speed and stability using classical adaptive filter theory. arXiv preprint arXiv:2002.10674.

26. Kohler, J., H. Daneshmand, A. Lucchi, T. Hofmann, M. Zhou, and K. Neymeyr (2019). Exponential convergence rates for batch normalization: The power of length-direction decoupling in non-convex optimization. In *AISTATS*.

27. Dukler, Y., Q. Gu, and G. Montúfar (2020). Optimization theory for relu neural networks trained with normalization layers. In *ICML*.

28. Nair, V. and G. E. Hinton (2010). Rectified linear units improve restricted boltzmann machines. In *ICML*.

29. Heo, B., S. Chun, S. J. Oh, D. Han, S. Yun, Y. Uh, and J. Ha (2021). Slowing down the weight norm increase in momentum-based optimizers. In *ICLR*.

30. Krogh, A. and J. A. Hertz (1992). A simple weight decay can improve generalization. In *NeurIPS*.
31. Van Laarhoven, T. (2017). L2 regularization versus batch and weight normalization. arXiv preprint arXiv:1706.05350.
32. Huang, L., X. Liu, B. Lang, and B. Li (2017). Projection based weight normalization for deep neural networks. arXiv preprint arXiv:1710.02338.
33. Zhang, G., C. Wang, B. Xu, and R. B. Grosse (2019). Three mechanisms of weight decay regularization. In *ICLR*.
34. Li, X., S. Chen, and J. Yang (2020). Understanding the disharmony between weight normalization family and weight decay. In *AAAI*.
35. Li, Z. and S. Arora (2020). An exponential learning rate schedule for batch normalized networks. In *ICLR*.
36. Desjardins, G., K. Simonyan, R. Pascanu, and k. kavukcuoglu (2015). Natural neural networks. In *NeurIPS*.
37. LeCun, Y., I. Kanter, and S. A. Solla (1990). Second order properties of error surfaces. In *NeurIPS*.
38. Martens, J. and R. Grosse (2015). Optimizing neural networks with kronecker-factored approximate curvature. In *ICML*.
39. Daneshmand, H., J. Kohler, F. Bach, T. Hofmann, and A. Lucchi (2020). Theoretical understanding of batch-normalization: A markov chain perspective. arXiv preprint arXiv:2003.01652.
40. Lubana, E. S., R. Dick, and H. Tanaka (2021). Beyond batchnorm: Towards a unified understanding of normalization in deep learning. In *NIPS*.
41. Santurkar, S., D. Tsipras, A. Ilyas, and A. Madry (2018). How does batch normalization help optimization? In *NeurIPS*.
42. Yao, Z., A. Gholami, K. Keutzer, and M. W. Mahoney (2020). PyHessian: Neural networks through the lens of the Hessian. In *2020 IEEE International Conference on Big Data (Big Data)*.
43. Ghorbani, B., S. Krishnan, and Y. Xiao (2019). An investigation into neural net optimization via Hessian eigenvalue density. In *ICML*.
44. Bjorck, J., C. Gomes, and B. Selman (2018). Understanding batch normalization. In *NeurIPS*.
45. Karakida, R., S. Akaho, and S.-i. Amari (2019). The normalization method for alleviating pathological sharpness in wide neural networks. In *NeurIPS*, pp. 6403–6413.
46. Mishkin, D. and J. Matas (2016). All you need is a good init. In *ICLR*.
47. Chen, G., P. Chen, Y. Shi, C. Hsieh, B. Liao, and S. Zhang (2019). Rethinking the usage of batch normalization and dropout in the training of deep neural networks. arXiv preprint arXiv:1905.05928.
48. Wei, M., J. Stokes, and D. J. Schwab (2019). Mean-field analysis of batch normalization. arXiv preprint arXiv:1903.02606.
49. Labatie, A. (2019). Characterizing well-behaved vs. pathological deep neural networks. In *ICML*, pp. 3611–3621.
50. Lee, J., J. Sohl-dickstein, J. Pennington, R. Novak, S. Schoenholz, and Y. Bahri (2018). Deep neural networks as gaussian processes. In *ICLR*.
51. Shekhovtsov, A. and B. Flach (2018b). Stochastic normalizations as bayesian learning. In *ACCV*.
52. Liang, S., Z. Huang, M. Liang, and H. Yang (2020). Instance enhancement batch normalization: an adaptive regulator of batch noise. In *AAAI*.
53. Huang, L., Y. Zhou, F. Zhu, L. Liu, and L. Shao (2019). Iterative normalization: Beyond standardization towards efficient whitening. In *CVPR*.
54. Luo, P., X. Wang, W. Shao, and Z. Peng (2019). Towards understanding regularization in batch normalization. In *ICLR*.
55. Huang, L., L. Zhao, Y. Zhou, F. Zhu, L. Liu, and L. Shao (2020). An investigation into the stochasticity of batch whitening. In *CVPR*.

56. Srivastava, N., G. Hinton, A. Krizhevsky, I. Sutskever, and R. Salakhutdinov (2014, January). Dropout: A simple way to prevent neural networks from overfitting. *J. Mach. Learn. Res. 15*(1), 1929–1958.

57. Li, X., S. Chen, X. Hu, and J. Yang (2019). Understanding the disharmony between dropout and batch normalization by variance shift. In *CVPR*.

58. Teye, M., H. Azizpour, and K. Smith (2018). Bayesian uncertainty estimation for batch normalized deep networks. In *ICML*.

59. Atanov, A., A. Ashukha, D. Molchanov, K. Neklyudov, and D. Vetrov (2018). Uncertainty estimation via stochastic batch normalization. In *ICLR Workshop*.

60. Nado, Z., S. Padhy, D. Sculley, A. D'Amour, B. Lakshminarayanan, and J. Snoek (2020). Evaluating prediction-time batch normalization for robustness under covariate shift. arXiv preprint arXiv:2006.10963.

61. Hoffer, E., I. Hubara, and D. Soudry (2017). Train longer, generalize better: closing the generalization gap in large batch training of neural networks. In *NeurIPS*.

62. Summers, C. and M. J. Dinneen (2020). Four things everyone should know to improve batch normalization. In *ICLR*.

63. Dimitriou, N. and O. Arandjelovic (2020). A new look at ghost normalization. arXiv preprint arXiv:2007.08554.

64. Huang, L., Y. Zhou, L. Liu, F. Zhu, and L. Shao (2021). Group whitening: Balancing learning efficiency and representational capacity. In *CVPR*.

65. Zhang, C., S. Bengio, M. Hardt, B. Recht, and O. Vinyals (2017). Understanding deep learning requires rethinking generalization. In *ICLR*.

66. Vapnik, V. N. (1999). An overview of statistical learning theory. *IEEE transactions on neural networks 10*(5), 988–999.

Normalization in Task-Specific Applications

10

As previously stated, normalization methods can be wrapped as general modules, which have been extensively integrated into various DNNs to stabilize and accelerate training, probably leading to improved generalization. For example, BN is an essential module in the state-of-the-art network architectures for computer vision (CV) tasks [1–6], and LN is an essential module in natural language processing (NLP) tasks [7–9]. In this chapter, we discuss the applications of normalization for particular tasks, in which normalization methods can effectively solve the key issues. Generally speaking, the main motivation of normalization for specific application is that the statistics calculated by normalization can represent specific domain information for visual tasks. For example, the statistics of a set of images sometimes can be used to represent domain information, that is, this set of images are sampled from the same situations. Also, the statistics of one image sometimes can represent the style of the image (Fig. 10.1). Thus, it is possible to learn domain invariant representation by aligning the distributions among different domains using normalization, for discriminative models. It is also possible to edit style information, by using the normalization operation (NOP) to remove style information, and using the normalization representation recovery (NRR) to add another style.

To be specific, this book mainly review the applications of normalization in domain adaptation, style transfer, training generative adversarial networks (GANs) and efficient deep models. However, we note that there also exist works exploring how to apply normalization to meta learning [10–12], reinforcement learning [13–15], unsupervised/self-supervised representation learning [16, 17], permutation-equivariant networks [18, 19], graph neural networks [20], ordinary differential equation (ODE) based networks [21], symmetric positive definite (SPD) neural networks [22], and guarding against adversarial attacks [23–25].

© The Author(s), under exclusive license to Springer Nature Switzerland AG 2022
L. Huang, *Normalization Techniques in Deep Learning*, Synthesis Lectures on Computer
Vision, https://doi.org/10.1007/978-3-031-14595-7_10

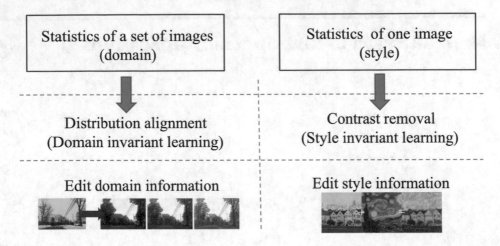

Fig. 10.1 The general idea of normalization for learning invariant representation and editing the distribution corresponding to the visual prior

10.1 Domain Adaptation

Machine learning algorithms trained on some given data (source domain) usually perform poorly when tested on data acquired under different settings (target domain). This is explained in domain adaptation as resulting from a shift between the distributions of the source and target domains. Most methods for domain adaptation thus aim to bridge the gap between these distributions. A typical way of achieving this is to align the distributions of the source and target domains based on the mini-batch/population statistics of BNs [26].

Algorithm 10.1 Adaptive Batch Normalization (AdaBN)

1: **for** neuron j in DNN **do**

2:      Concatenate neuron responses on all images of target domain $t : \mathbf{x}_j = [\ldots, x_j(m), \ldots]$

3:      Compute the mean and variance of the target domain: $\mu_j^t = \mathbb{E}(\mathbf{x}_j^t), \sigma_j^t = \sqrt{Var(\mathbf{x}_j^t)}$.

4: **end for**

5: **for** neuron j in DNN, testing image m in target domain **do**

6:      Compute BN output $y_j(m) := \gamma_j \dfrac{(x_j(m) - \mu_j^t)}{\sigma_j^t} + \beta_j$

7: **end for**

Li et al. [26] are the first to investigate BNs' statistics in domain adaptation. They observe that the statistics of BN layer contain the traits of the data domain. Besides, both shallow layers and deep layers of the DNN are influenced by domain shift, and domain adaptation by manipulating the output layer alone is not enough. Based on these observations, they propose adaptive batch normalization (AdaBN), where the BN statistics for the source domain are calculated during training, and then those for the target domain are modulated during testing (see Algorithm 10.1). AdaBN enables domain-invariant features to be learnt without requiring additional loss terms and the extra associated parameters. It can directly adapt to the target domain, by only re-calculating the population statistics of BN.

The hypothesis behind AdaBN is that the domain-invariant information is stored in the weight matrix of each layer, while the domain-specific information is represented by the statistics of the BN layer. However, this hypothesis may not always hold because the target domain is not exploited at the training stage. As a result, it is difficult to ensure that the statistics of the BN layers in the source and target domains correspond to their domain-specific information.

One way to overcome this limitation is to couple the network parameters for both target and source samples in the training stage, which has been the main research focus of several follow-up works inspired by AdaBN. Carlucci et al. [27] proposed automatic domain alignment layers (AutoDIAL), which are embedded in different levels of the deep architecture to align the learned source and target feature distributions to a canonical one. AutoDIAL exploits the source and target features during the training stage, in which an extra parameter is involved in each BN layer as the trade-off between the source and target domains. Chang et al. [28] further proposed domain-specific batch normalization (DSBN), where multiple branches of BN are used, each of which is exclusively in charge of a single domain. DSBN is different to the original BN, where the data from different domain is normalized with the same BN. DSBN learns domain-specific properties using only the estimated population statistics of BN and learns domain-invariant representations with the other parameters in the network. This method effectively separates domain-specific information for unsupervised domain adaptation. Similar ideas have also been exploited in unsupervised adversarial domain adaptation in the context of semantic scene segmentation [29] and adversarial examples for improving image recognition [30]. Roy et al. [31] further generalized DSBN by a domain-specific whitening transform (DWT), where the source and target data distributions are aligned using their covariance matrices. DWT layers "whiten" the source and the target features and project them into a common spherical distribution. Notably, DWT generalizes previous BN-based DA methods which do not consider inter-feature correlations and rely only on feature standardization. DWT further consider inter-feature correlations and provide a stronger alignment.

Wang et al. [32] proposed transferable normalization (TransNorm), which also calculates the statistics of inputs from the source and target domains separately, while computing the channel transferability simultaneously. The normalized features then go through channel-adaptive mechanisms to re-weight the channels according to their transferability.

10.1.1 Domain Generalization

The previous work we introduced focus on the unsupervised domain-adaptation task, where the unlabeled target domain data can be available for adaptation. Here, we introduce normalization's application in domain generalization task, where examples in the target domain cannot be accessed during training. This task is considered to be more challenging than unsupervised domain adaptation.

Seo et al. [33] proposed domain-specific optimized normalization to learn domain-invariant representation for domain generalization. They also use different BN for different domains during training. Besides population statistics, they also exploited the affine transform (Eq. 4.3) of BN to represent the domain-specific information in their proposed domain-specific optimized normalization (DSON). Moreover, DSON normalizes the activations by a weighted average of multiple normalization statistics (typically BN and IN), and keeps track of the normalization statistics of each normalization type if necessary, for each domain.

Choi et al. [34] present an instance selective whitening loss that alleviates the limitations of the existing whitening transformation for domain generalization. Instance selective whitening selectively removing information that causes a domain shift while maintaining a discriminative power of feature within DNNs. The proposed method does not rely on an explicit closed-form whitening transformation, but implicitly encourage the networks to learn such a whitening transformation through the proposed loss function. They selectively remove only those feature covariances that respond sensitively to photometric augmentation such as color transformation.

Segù et al. [35] propose batch normalization embedding for deep domain generalization. They propose a multi-source domain alignment layer. This layer collects domain-specific population statistics and compute instance statistics for test samples. After training, the population and instance statistics map respectively the source domains and the test samples into a latent space, where domain similarity can be measured by distances between embedding vectors.

10.1.2 Robust Deep Learning Under Covariate Shift

The idea of normalization from domain adaption can be applied for corruption robust problems, if different corruptions can be viewed as different domains. There are some works investigating normalization for improving corruption robust under covariate shift. Similar to the adaptive batch normalization used in domain adaption, some works address to estimate the target statistics during test to improve the performance for corruption. For example, Schneider et al. [36] propose to exploit test images to estimate the BN's statistics, and they also consider combining the statistics of training images if the number of test images are small. Benz et al. [37] interpret corruption robustness as a domain shift and propose to rectify BN statistics (mean and variance) for improving model robustness using the test images. This is motivated by perceiving the shift from the clean domain to the corruption domain as a style shift that is represented by the BN statistics. Nado et al. [38] consider a prediction scenario, where the mini-batch data can be *i.i.d* obtained during prediction stages. Under this scenario, they propose prediction-time batch normalization that recomputes the BN statistics for each test batch, rather than uses the population statistics calculated during training. This strategy can effectively improve the corruption robust. This prediction-time batch normalization is implemented as the standard "training mode" of BN in Pytorch or Tensorflow.

Wang et al. [39] also propose fully test-time adaptation (Tent) for domain adaption or image corruption. They observe that prediction with lower entropy have lower error rates on corrupted CIFAR-100-C datasets and certainty can serve as supervision during testing. They thus use entropy minimization on test data as an optimization objective. Tent modulates features during testing by estimating normalization statistics μ and σ, and optimizing affine parameters γ and β. The normalization statistics and affine parameters are updated on target data without use of source data. It is efficient to adapt γ and β in practice, because they make up less than 1% of model parameters

Ishii and Sugiyama [40] also propose similar idea for source-free domain adaptation. They jointly minimize both the BN-statistics matching loss and information maximization loss to fine-tune the encoder. As to the BN-statistics matching loss, The distribution of unobservable source features are summarized and stored as the BN statistics in the BN layer, and the loss explicitly evaluates the discrepancy between source and target feature distributions based on those statistics. Therefore, minimizing this loss can also align the distribution of features extracted by the fine-tune encoder between source and target domain.

The idea of learning domain-invariant representation by using BN is also used for defending multiple adversarial examples. Liu et al. [41] observe that different types of adversarial perturbations induce different statistical properties that can be separated and characterized by the statistics of BN. They thus proposed gated BN (GBN) to adversarially train a perturbation-invariant predictor for defending multiple perturbation types. GBN consists of a multibranch BN layer and a gated sub-network. Each BN branch in GBN is in charge of one perturbation type to ensure that the normalized output is aligned for further learning perturbation-invariant representation. Meanwhile, the gated sub-network is designed to separate model inputs added with different perturbations. By this design, GBN can well defending multiple seen adversarial perturbations, and even unseen adversarial perturbations.

There are some works exploiting the idea of BN in domain adaption for person re-identification (ReID). Zhuang et al. [42] investigate the distribution with respect to each camera for person re-identification tasks. They observe that different cameras show remarkable differences in distributions of visual data. Based on this observation, They view each camera as a "domain", and emphasize the importance of aligning the distribution of all cameras. They propose camera based batch normalization (CBN), which looks like the domain-specific BN, but used for person re-identification tasks. In training, CBN disassembles each mini-batch and standardizes the corresponding input according to its camera labels. In testing, CBN utilizes few samples to approximate the BN statistics of every testing camera and standardizes the input to the training distribution. CBN facilitates the generalization and transfer ability of ReID models across different scenarios and makes better use of intra-camera annotations. Similar idea is also used for the visible modality and infrared modality in person re-identification scenario. Li et al. [43] view each modality as a "domain", and proposes modality batch normalization. It also aims to align the distribution of different modality for better learning invariant content representation.

There is also work that exploit the idea of BN in domain adaption for Stereo matching. Zhang et al. [44] propose domain-invariant stereo match network, which aims at generalizing well to unseen scenes. They propose domain invariant normalization. It normalizes features along the spatial axis (height and width) to induce style-invariant representations similar to instance normalization. Besides, the L2-Norm based scaling is also used to normalize features along the channel axis for each spatial position, like pixel normalization.

10.1.3 Learning Universal Representations

The idea of applying BN in domain adaptation can be further extended to the learning of universal representations [45], by constructing neural networks that work simultaneously in many domains. To achieve this, the networks need to learn to share common visual structures where no obvious commonality exists. Universal representations cannot only benefit domain adaptation but also contribute to multi-task learning, which aims to learn multiple tasks simultaneously in the same data domain.

Bilen and Vedaldi [45] advocate to learn universal image representations using (1) the convolutional kernels to extract domain-agnostic information and (2) the BN layers to transform the internal representations to the relevant target domains. Wesley Putra Data et al. [46] exploited BN layers to learn discriminate visual classes, while other layers (e.g., convolutional layers) are used to learn the universal representation. They also provided a way to interpolate BN layers to solve new tasks. Li and Vasconcelos [47] proposed covariance normalization (CovNorm) for multi-domain learning, which provides efficient solutions to several tasks defined in different domains. Mudrakarta et al. [48] propose a learning paradigm for multi-task learning, in which each task carries its own model patches. These model patches refer to a small set of parameters that along with a shared set of parameters constitute the model for that task. For example, they use both BN's population statistics (mean and variance) and the affine parameters for each task.

10.2   Image Style Transfer

Image style transfer is an important image editing task that enables the creation of new artistic works [49, 50]. Image style transfer algorithms aim to generate a stylized image that has similar content and style to the given images. The key challenge in this task is to extract effective representations that can disentangle the style from the content. The seminal work by Gatys et al. [51] showed that the covariance/Gram matrix of the layer activations, extracted by a trained DNN, has a remarkable capacity for capturing visual styles. This provides a feasible solution to matching the styles between images by minimizing Gram matrix based losses, pioneering the way for style transfer.

A key advantage of applying normalization to style transfer is that the normalization operation (NOP) can remove the style information (e.g., whitening can ensure the covariance matrix to be an identity matrix), while the normalization representation recovery (NRR), in contrast, introduces it. In other words, the style information is intuitively 'editable' by normalization [52, 53]. In a seminal work, Ulyanov et al. [54] proposed instance normalization (IN) to remove instance-specific contrast information (style) from the content image. Since then, IN has been a basic module for image style transfer tasks.

Dumoulin et al. [55] proposed conditional instance normalization (CIN), an efficient solution to integrating multiple styles. Specifically, multiple distinct styles are captured by a single network, by encoding the style information in the affine parameters (Eq. 4.3) of IN layers, after which each style can be selectively applied to a target image. Huang and Belongie [56] proposed adaptive instance normalization (AdaIN), where the activations of content images are standardized by their statistics, and the affine parameters (β and γ) come from the statistics of style activations. AdaIN transfers the channel-wise mean and variance feature statistics between content and style feature activations. AdaIN can also work well in text effect transfer, which aims at learning the visual effects while maintaining the text content [57]. Rather than manually defining how to compute the affine parameters so as to align the mean and variance between content and style features, dynamic instance normalization (DIN), introduced by Jing et al. [58], deals with arbitrary style transfer by encoding a style image into learnable convolution parameters, upon which the content image is stylized.

To address the limitations of AdaIN in only trying to match up the variances of the stylized image and the style image feature, Li et al. [49] further proposed whitening and coloring transformations (WCT) to match up the covariance matrix. This shares a similar spirit to the optimization of the Gram matrix based cost in neural style transfer [49, 59]. Some methods [60] also seek to provide a good trade-off between AdaIN (which enjoys higher computational efficiency) and WCT (which synthesizes images visually closer to a given style).

10.2.1 Image Translation

In computer vision, image translation can be viewed as a more general case of image style transfer. Given an image in the source domain, the aim is to learn the conditional distribution of the corresponding images in the target domain. This includes, but is not limited to, the following tasks: super-resolution, colorization, inpainting, and attribute transfer. Similar to style transfer, AdaIN is also an essential tool for image translation used in, for example, multimodal unsupervised image-to-image translation (MUIT) [61]. Note that the affine parameters of AdaIN in [61] are produced by a learned network, instead of computed from statistics of a pretrained network as in [56]. Apart from IN, Cho et al. [62] proposed the group-wise deep whitening-and-coloring transformation (GDWCT) by matching higher-order statistics, such as covariance, for image-to-image translation tasks. Moreover, since the whitening/coloring transformation can be considered a 1×1 convolution, Cho et al. [63] further proposed adaptive convolution-based normalization (AdaCoN) to inject the target style into a given image, for unsupervised image-to-image translation. AdaCoN first performs standardization locally on each subregion of an input activation map (similar to the local normalization shown in Sect. 4.1) and then applies an adaptive convolution, where the convolution filter weights are dynamically estimated using the encoded style representation. Besides, Yu et al. [64] proposed region normalization (RN) for image inpainting network training. RN divides spatial pixels into different regions according to the input mask and standardizes the activations in each region. Wang [65] introduced attentive normalization (AN) for conditional image generation, which is an extension of instance normalization [54]. AN divides the feature maps into different regions based on their semantics, and then separately normalizes and denormalizes the feature points in the same region.

10.3   Training GANs

GANs [66] can be regarded as a general framework to produce a model distribution that mimics a given target distribution. A GAN consists of a generator, which produces the model distribution, and a discriminator, which distinguishes the model distribution from the target. From this perspective, the ultimate goal when training GANs shares a similar spirit to model training in the domain adaptation task. The main difference lies in that GANs try to reduce the distance between different distributions, while domain adaptation models attempt to close the gap between different domains. Therefore, the techniques that apply BN to domain adaptation, as discussed in Sect. 10.1, may work for GANs as well. For example, combining samples form different domains in a batch for BN may harm the generalization in domain adaptation, and this also applies for the training of GANs [67, 68]. It is better to use the different BN modules for generated examples and real examples in discriminator, rather than the single one, which is similar to the domain-specific BN. The main reason is that the generated examples and real examples are easily distinguished by the discriminator in the initial states , if use only one BN. This leads to that the discriminator provides few gradient information to align the distributions.

   One persisting challenge in training GANs is the performance control of the discriminator and the learning pace control between the discriminator and generator [69]. The density ratio estimated by the discriminator is often inaccurate and unstable during training, and the generator may fail to learn the structure of the target distribution. One way to remedy this issue is to impose constraints on the discriminator [70]. For instance, Xiang and Li [71] leveraged weight normalization to effectively improve the training performance of GANs. Miyato et al. [69] proposed spectral normalization (SN), which enforces Lipschitz continuity on the discriminator by normalizing its parameters with the spectral norm estimated by power iteration. Since then, SN has become an important technique in training GANs [69, 72, 73]. Zhang et al. [74] further found that employing SN in the generator improves the stability, allowing for fewer training steps for the discriminator per iteration. Another important constraint in training GANs is the orthogonality [73, 75–77]. Reference [73] found that applying orthogonal regularization to the generator renders it amenable to a simple 'truncation trick', allowing fine control over the trade-off between sample fidelity and variety by reducing the variance of the generator input. Huang et al. [75] proposed orthogonalization by Newton's iteration, which can effectively control the orthogonality of the weight matrix, and interpolate between spectral normalization and full orthogonalization by altering the iteration number.

As discussed in Sect. 10.2 for style transfer, the NRR operation of activation normalization can also be used as the side information for GANs, under the scenario of conditional GANs (cGANs) [78]. cGANs have shown advancements in class conditional image generation [79], image generation from text [80, 81], and image-to-image translation [82].

de Vries et al. [83] proposed conditional batch normalization (CBN), which injects a linguistic input (e.g., a question in a VQA task) into the affine parameters of BN. This shares a similar spirit to the conditional instance normalization for style transfer, and has been extensively explored in [73, 74, 84, 85]. Karras et al. [86] proposed a style-based generator architecture for GANs, where the style information is embedded into the affine parameters of AdaIN [56]. Note that the style comes from the latent vector instead of an example image, enabling the model to work without external information. Similarly, Chen et al. [87] proposed a more general self-modulation based on CBN, where the affine parameters can also be generated by the generator's own input or provided by external information.

10.4 Efficient Deep Models

In real-world applications, it is essential to consider the efficiency of an algorithm in addition to its effectiveness due to the often limited computational resources (such as in smartphones). As such, there is also an active line of research exploiting normalization techniques (e.g., BN) to develop efficient DNNs based on network slimming or quantization. In network slimming, the general idea is to exploit the channel-wise scale parameter $\gamma \in \mathbb{R}^d$ of BN, considering that each scale γ_i corresponds to a specific convolutional channel (or a neuron in a fully connected layer) [88]. Liu et al. For example, [88] proposed to identify and prune insignificant channels (or neurons) based on the scale parameter in BN layers, which are imposed by L^1 regularization for sparsity. Ye et al. [89] also adopted a similar idea, and developed a new algorithmic approach and rescaling trick to improve the robustness and speed of optimization. Li et al. [90] proposed an efficient evaluation component based on adaptive batch normalization [26], which has a strong correlation between different pruned DNN structures and their final settled accuracy.

Yu et al. [91] trained a slimmable network with a new variant of BN, namely switchable batch normalization (SBN), for the networks executable at different widths. SBN privatizes BN for different switches of a slimmable network, and each individual BN has independently accumulated feature statistics. SBN can thus be used as a general solution to obtain a good trade-off between accuracy and latency on the fly. As a complement to BN that normalizes the final summation of the weighted inputs, Luo et al. [92] proposed fine-grained batch normalization (FBN) to build light-weight networks, where FBN normalizes the intermediate state of the summation.

Network quantization is another essential technique in building efficient DNNs. This challenging task can also be tackled using normalization algorithms like BN. For instance, Banner et al. [93] proposed range batch normalization (RBN) for quantized networks, normalizing activations according to the range of the activation distribution. RBN avoids the sum of squares, square-root and reciprocal operations and is more friendly for low-precise training [94]. Lin et al. [95] proposed to quantize BN in model deployment by converting the two floating points affine transformations to a fixed-point operation with shared quantized scale. Ardakani et al. [96] employed BN to train binarized/ternarized LSTMs, and achieved state-of-the-art performance in network quantization. Hou et al. [97] further studied and compared the quantized LSTMs with WN, LN and BN. They showed that these normalization methods make the gradient invariant to weight scaling, thus alleviating the problem of having a potentially large weight norm increase due to quantization. Sari et al. [98] analyzed how the centering and scaling operations in BN affect the training of binary neural networks.

References

1. Russakovsky, O., J. Deng, H. Su, J. Krause, S. Satheesh, S. Ma, Z. Huang, A. Karpathy, A. Khosla, M. Bernstein, et al. (2015). Imagenet large scale visual recognition challenge. *International journal of computer vision 115*(3), 211–252.
2. He, K., X. Zhang, S. Ren, and J. Sun (2016a). Deep residual learning for image recognition. In *CVPR*.
3. Zagoruyko, S. and N. Komodakis (2016). Wide residual networks. In *BMVC*.
4. Szegedy, C., V. Vanhoucke, S. Ioffe, J. Shlens, and Z. Wojna (2016). Rethinking the inception architecture for computer vision. In *CVPR*.
5. Huang, G., Z. Liu, and K. Q. Weinberger (2017). Densely connected convolutional networks. In *CVPR*.
6. Xie, S., R. B. Girshick, P. Dollár, Z. Tu, and K. He (2017). Aggregated residual transformations for deep neural networks. In *CVPR*.
7. Vaswani, A., N. Shazeer, N. Parmar, J. Uszkoreit, L. Jones, A. N. Gomez, L. Kaiser, and I. Polosukhin (2017). Attention is all you need. In *NeurIPS*.
8. Yu, A. W., D. Dohan, M.-T. Luong, R. Zhao, K. Chen, M. Norouzi, and Q. V. Le (2018). Qanet: Combining local convolution with global self-attention for reading comprehension. In *ICLR*.

9. Xu, J., X. Sun, Z. Zhang, G. Zhao, and J. Lin (2019). Understanding and improving layer normalization. In *NeurIPS*.
10. Nichol, A., J. Achiam, and J. Schulman (2018). On first-order meta-learning algorithms. arXiv preprint arXiv:1803.02999.
11. Gordon, J., J. Bronskill, M. Bauer, S. Nowozin, and R. Turner (2019). Meta-learning probabilistic inference for prediction. In *ICLR*.
12. Bronskill, J., J. Gordon, J. Requeima, S. Nowozin, and R. E. Turner (2020a). Tasknorm: Rethinking batch normalization for meta-learning. In *ICML*.
13. van Hasselt, H. P., A. Guez, M. Hessel, V. Mnih, and D. Silver (2016). Learning values across many orders of magnitude. In *NeurIPS*.
14. Bhatt, A., M. Argus, A. Amiranashvili, and T. Brox (2019). Crossnorm: Normalization for off-policy TD reinforcement learning. arXiv preprint arXiv:1902.05605.
15. Wang, C., Y. Wu, Q. Vuong, and K. Ross (2020). Striving for simplicity and performance in off-policy DRL: Output normalization and non-uniform sampling. In *ICML*.
16. He, K., H. Fan, Y. Wu, S. Xie, and R. Girshick (2020). Momentum contrast for unsupervised visual representation learning. In *CVPR*.
17. Taha Kocyigit, M., L. Sevilla-Lara, T. M. Hospedales, and H. Bilen (2020). Unsupervised batch normalization. In *CVPR Workshops*.
18. Moo Yi, K., E. Trulls, Y. Ono, V. Lepetit, M. Salzmann, and P. Fua (2018). Learning to find good correspondences. In *CVPR*.
19. Sun, W., W. Jiang, E. Trulls, A. Tagliasacchi, and K. M. Yi (2020). Attentive context normalization for robust permutation-equivariant learning. In *CVPR*.
20. Cai, T., S. Luo, K. Xu, D. He, T.-y. Liu, and L. Wang (2020). Graphnorm: A principled approach to accelerating graph neural network training. arXiv preprint arXiv:2009.03294.
21. Gusak, J., L. Markeeva, T. Daulbaev, A. Katrutsa, A. Cichocki, and I. Oseledets (2020). Towards understanding normalization in neural odes. arXiv preprint arXiv:2004.09222.
22. Brooks, D., O. Schwander, F. Barbaresco, J.-Y. Schneider, and M. Cord (2019). Riemannian batch normalization for spd neural networks. In *NeurIPS*, pp. 15463–15474.
23. Galloway, A., A. Golubeva, T. Tanay, M. Moussa, and G. W. Taylor (2019). Batch normalization is a cause of adversarial vulnerability. arXiv preprint arXiv:1905.02161.
24. Awais, M., F. Shamshad, and S.-H. Bae (2020). Towards an adversarially robust normalization approach. arXiv preprint arXiv:2006.11007.
25. Xie, C. and A. Yuille (2020). Intriguing properties of adversarial training at scale. In *ICLR*.
26. Li, Y., N. Wang, J. Shi, J. Liu, and X. Hou (2016). Revisiting batch normalization for practical domain adaptation. arXiv preprint arXiv:1603.04779.
27. Carlucci, F. M., L. Porzi, B. Caputo, E. Ricci, and S. R. Bulo (2017). Autodial: Automatic domain alignment layers. In *ICCV*.
28. Chang, W., T. You, S. Seo, S. Kwak, and B. Han (2019). Domain-specific batch normalization for unsupervised domain adaptation. In *CVPR*.
29. Romijnders, R., P. Meletis, and G. Dubbelman (2019). A domain agnostic normalization layer for unsupervised adversarial domain adaptation. In *WACV*.
30. Xie, C., M. Tan, B. Gong, J. Wang, A. Yuille, and Q. V. Le (2020). Adversarial examples improve image recognition. In *CVPR*.
31. Roy, S., A. Siarohin, E. Sangineto, S. R. Bulo, N. Sebe, and E. Ricci (2019). Unsupervised domain adaptation using feature-whitening and consensus loss. In *CVPR*.
32. Wang, X., Y. Jin, M. Long, J. Wang, and M. I. Jordan (2019). Transferable normalization: Towards improving transferability of deep neural networks. In *NeurIPS*.
33. Seo, S., Y. Suh, D. Kim, J. Han, and B. Han (2020). Learning to optimize domain specific normalization for domain generalization. In *ECCV*.

34. Choi, S., S. Jung, H. Yun, J. T. Kim, S. Kim, and J. Choo (2021). Robustnet: Improving domain generalization in urban-scene segmentation via instance selective whitening. In *Proceedings of the IEEE/CVF Conference on Computer Vision and Pattern Recognition*, pp. 11580–11590.

35. Segù, M., A. Tonioni, and F. Tombari (2020). Batch normalization embeddings for deep domain generalization. *CoRR abs/2011.12672*.

36. Schneider, S., E. Rusak, L. Eck, O. Bringmann, W. Brendel, and M. Bethge (2020). Improving robustness against common corruptions by covariate shift adaptation. In *Advances in Neural Information Processing Systems*.

37. Benz, P., C. Zhang, A. Karjauv, and I. S. Kweon (2020). Revisiting batch normalization for improving corruption robustness. *CoRR abs/2010.03630*.

38. Nado, Z., S. Padhy, D. Sculley, A. D'Amour, B. Lakshminarayanan, and J. Snoek (2020). Evaluating prediction-time batch normalization for robustness under covariate shift. arXiv preprint arXiv:2006.10963.

39. Wang, D., E. Shelhamer, S. Liu, B. Olshausen, and T. Darrell (2021). Tent: Fully test-time adaptation by entropy minimization. In *International Conference on Learning Representations*.

40. Ishii, M. and M. Sugiyama (2021). Source-free domain adaptation via distributional alignment by matching batch normalization statistics. *CoRR abs/2101.10842*.

41. Liu, A., S. Tang, X. Liu, X. Chen, L. Huang, Z. Tu, D. Song, and D. Tao (2020). Towards defending multiple adversarial perturbations via gated batch normalization. *CoRR abs/2012.01654*.

42. Zhuang, Z., L. Wei, L. Xie, T. Zhang, H. Zhang, H. Wu, H. Ai, and Q. Tian (2020). Rethinking the distribution gap of person re-identification with camera-based batch normalization. In *European Conference on Computer Vision*, pp. 140–157. Springer.

43. Li, W., Q. Ke, W. Chen, and Y. Zhou (2021). Bridging the distribution gap of visible-infrared person re-identification with modality batch normalization. *CoRR abs/2103.04778*.

44. Zhang, F., X. Qi, R. Yang, V. Prisacariu, B. Wah, and P. Torr (2020). Domain-invariant stereo matching networks. In *Europe Conference on Computer Vision (ECCV)*.

45. Bilen, H. and A. Vedaldi (2017). Universal representations: The missing link between faces, text, planktons, and cat breeds. arXiv preprint arXiv:1701.07275.

46. Wesley Putra Data, G., K. Ngu, D. William Murray, and V. Adrian Prisacariu (2018). Interpolating convolutional neural networks using batch normalization. In *ECCV*.

47. Li, Y. and N. Vasconcelos (2019). Efficient multi-domain learning by covariance normalization. In *CVPR*.

48. Mudrakarta, P. K., M. Sandler, A. Zhmoginov, and A. Howard (2019). K for the price of 1: Parameter efficient multi-task and transfer learning. In *International Conference on Learning Representations*.

49. Li, Y., C. Fang, J. Yang, Z. Wang, X. Lu, and M.-H. Yang (2017). Universal style transfer via feature transforms. In *NeurIPS*.

50. Jing, Y., Y. Yang, Z. Feng, J. Ye, Y. Yu, and M. Song (2019). Neural style transfer: A review. *IEEE transactions on visualization and computer graphics*.

51. Gatys, L. A., A. S. Ecker, and M. Bethge (2016). Image style transfer using convolutional neural networks. In *CVPR*.

52. Li, B., F. Wu, K. Q. Weinberger, and S. Belongie (2019). Positional normalization. In *NeurIPS*.

53. Li, B., F. Wu, S.-N. Lim, S. Belongie, and K. Q. Weinberger (2020). On feature normalization and data augmentation. arXiv preprint arXiv:2002.11102.

54. Ulyanov, D., A. Vedaldi, and V. S. Lempitsky (2016). Instance normalization: The missing ingredient for fast stylization. arXiv preprint arXiv:1607.08022.

55. Dumoulin, V., J. Shlens, and M. Kudlur (2017). A learned representation for artistic style. In *ICLR*.

56. Huang, X. and S. Belongie (2017). Arbitrary style transfer in real-time with adaptive instance normalization. In *ICCV*.

57. Li, W., Y. He, Y. Qi, Z. Li, and Y. Tang (2020). Fet-GAN: Font and effect transfer via k-shot adaptive instance normalization. In *AAAI*.

58. Jing, Y., X. Liu, Y. Ding, X. Wang, E. Ding, M. Song, and S. Wen (2020). Dynamic instance normalization for arbitrary style transfer. In *AAAI*.

59. Chiu, T.-Y. (2019). Understanding generalized whitening and coloring transform for universal style transfer. In *ICCV*.

60. Sheng, L., Z. Lin, J. Shao, and X. Wang (2018). Avatar-net: Multi-scale zero-shot style transfer by feature decoration. In *CVPR*.

61. Huang, X., M. Liu, S. J. Belongie, and J. Kautz (2018). Multimodal unsupervised image-to-image translation. In *ECCV*.

62. Cho, W., S. Choi, D. K. Park, I. Shin, and J. Choo (2019). Image-to-image translation via group-wise deep whitening-and-coloring transformation. In *CVPR*.

63. Cho, W., K. Kim, E. Kim, H. J. Kim, and J. Choo (2019). Unpaired image translation via adaptive convolution-based normalization. arXiv preprint arXiv:1911.13271.

64. Yu, T., Z. Guo, X. Jin, S. Wu, Z. Chen, W. Li, Z. Zhang, and S. Liu (2020). Region normalization for image inpainting. In *AAAI*.

65. Wang, Y., Y.-C. Chen, X. Zhang, J. Sun, and J. Jia (2020). Attentive normalization for conditional image generation. In *CVPR*.

66. Goodfellow, I., J. Pouget-Abadie, M. Mirza, B. Xu, D. Warde-Farley, S. Ozair, A. Courville, and Y. Bengio (2014). Generative adversarial nets. In *NeurIPS*.

67. Radford, A., L. Metz, and S. Chintala (2015). Unsupervised representation learning with deep convolutional generative adversarial networks. arXiv preprint arXiv:1511.06434.

68. Salimans, T., I. Goodfellow, W. Zaremba, V. Cheung, A. Radford, X. Chen, and X. Chen (2016). Improved techniques for training GANs. In *NeurIPS*.

69. Miyato, T., T. Kataoka, M. Koyama, and Y. Yoshida (2018). Spectral normalization for generative adversarial networks. In *ICLR*.

70. Arjovsky, M., S. Chintala, and L. Bottou (2017). Wasserstein GAN. arXiv preprint arXiv:1701.07875.

71. Xiang, S. and H. Li (2017). On the effects of batch and weight normalization in generative adversarial networks. arXiv preprint arXiv:1704.03971.

72. Kurach, K., M. Lučić, X. Zhai, M. Michalski, and S. Gelly (2019). A large-scale study on regularization and normalization in GANs. In *ICML*.

73. Brock, A., J. Donahue, and K. Simonyan (2019). Large scale GAN training for high fidelity natural image synthesis. In *ICLR*.

74. Zhang, H., I. Goodfellow, D. Metaxas, and A. Odena (2019). Self-attention generative adversarial networks. In *ICML*.

75. Huang, L., L. Liu, F. Zhu, D. Wan, Z. Yuan, B. Li, and L. Shao (2020). Controllable orthogonalization in training DNNs. In *CVPR*.

76. Liu, B., Y. Zhu, Z. Fu, G. de Melo, and A. Elgammal (2020). OoGAN: Disentangling GAN with one-hot sampling and orthogonal regularization. In *AAAI*.

77. Müller, J., R. Klein, and M. Weinmann (2019). Orthogonal wasserstein GANs. arXiv preprint arXiv:1911.13060.

78. Mirza, M. and S. Osindero (2014). Conditional generative adversarial nets. arXiv preprint arXiv:1411.1784.

79. Odena, A., C. Olah, and J. Shlens (2017). Conditional image synthesis with auxiliary classifier GANs. In *ICML*.

80. Reed, S., Z. Akata, X. Yan, L. Logeswaran, B. Schiele, and H. Lee (2016). Generative adversarial text to image synthesis. In *ICML*.

81. Zhang, H., T. Xu, H. Li, S. Zhang, X. Wang, X. Huang, and D. Metaxas (2017). StackGAN: Text to photo-realistic image synthesis with stacked generative adversarial networks. In *ICCV*.

82. Zhu, J.-Y., T. Park, P. Isola, and A. A. Efros (2017). Unpaired image-to-image translation using cycle-consistent adversarial networks. In *ICCV*.

83. de Vries, H., F. Strub, J. Mary, H. Larochelle, O. Pietquin, and A. C. Courville (2017). Modulating early visual processing by language. In *NeurIPS*, pp. 6594–6604.

84. Miyato, T. and M. Koyama (2018). cGANs with projection discriminator. In *ICLR*.

85. Michalski, V., V. S. Voleti, S. E. Kahou, A. Ortiz, P. Vincent, C. Pal, and D. Precup (2019). An empirical study of batch normalization and group normalization in conditional computation. arXiv preprint arXiv:1908.00061.

86. Karras, T., S. Laine, and T. Aila (2019). A style-based generator architecture for generative adversarial networks. In *CVPR*.

87. Chen, T., M. Lucic, N. Houlsby, and S. Gelly (2019). On self modulation for generative adversarial networks. In *ICLR*.

88. Liu, Z., J. Li, Z. Shen, G. Huang, S. Yan, and C. Zhang (2017). Learning efficient convolutional networks through network slimming. In *ICCV*, pp. 2755–2763.

89. Ye, J., X. Lu, Z. Lin, and J. Z. Wang (2018). Rethinking the smaller-norm-less-informative assumption in channel pruning of convolution layers. In *ICLR*.

90. Li, B., B. Wu, J. Su, G. Wang, and L. Lin (2020). Eagleeye: Fast sub-net evaluation for efficient neural network pruning. In *ECCV*.

91. Yu, J., L. Yang, N. Xu, J. Yang, and T. Huang (2019). Slimmable neural networks. In *ICLR*.

92. Luo, C., J. Zhan, L. Wang, and W. Gao (2020). Finet: Using fine-grained batch normalization to train light-weight neural networks. arXiv preprint arXiv:2005.06828.

93. Banner, R., I. Hubara, E. Hoffer, and D. Soudry (2018). Scalable methods for 8-bit training of neural networks. In *NeurIPS*.

94. Graham, B. (2017). Low-precision batch-normalized activations. arXiv preprint arXiv:1702.08231.

95. Lin, D., P. Sun, G. Xie, S. Zhou, and Z. Zhang (2020). Optimal quantization for batch normalization in neural network deployments and beyond. arXiv preprint arXiv:2008.13128.

96. Ardakani, A., Z. Ji, S. C. Smithson, B. H. Meyer, and W. J. Gross (2019). Learning recurrent binary/ternary weights. In *ICLR*.

97. Hou, L., J. Zhu, J. Kwok, F. Gao, T. Qin, and T.-Y. Liu (2019). Normalization helps training of quantized LSTM. In *NeurIPS*.

98. Sari, E., M. Belbahri, and V. P. Nia (2019). How does batch normalization help binary training. arXiv preprint arXiv:1909.09139.

Summary and Discussion

11

In this book, we have provided a research landscape for normalization techniques, covering methods, analyses and applications. We believe that our work can provide valuable guidelines for selecting normalization techniques to use in training DNNs. With the help of these guidelines, it will be possible to design new normalization methods tailored to specific tasks (by the choice of NAP) or improve the trade-off between efficiency and performance (by the choice of NOP). We leave the following open problems for discussion.

Theoretical Perspective: While the practical success of DNNs is indisputable, their theoretical analysis is still limited. Despite the recent progress of deep learning in terms of representation [1], optimization [2] and generalization [3], the networks investigated theoretically are usually different from those used in practice [4]. One clear example is that, while normalization techniques are ubiquitously used in the current state-of-the-art architectures, the theoretical analyses for DNNs usually rule out them.

In fact, the methods commonly used for normalizing activations (e.g., BN, LN) often conflict with current theoretical analyses. For instance, in the representation of DNNs, one important strategy is to analyze the number of linear regions, where the expressivity of a DNN with rectifier nonlinearity can be quantified by the maximal number of linear regions it can separate its input space into [1, 5]. However, this generally does not hold if BN/LN are introduced, since they create nonlinearity, causing the theoretical assumptions to no longer be met. It is thus important to further investigate how BN/LN affect a model's representation capacity. As for optimization, most analyses require the input data to be independent, such that the stochastic/mini-batch gradient is an unbiased estimator of the true gradient over the dataset. However, BN typically does not fit this data-independent assumption, and its optimization usually depends on the sampling strategy as well as the mini-batch size [6]. There is thus a need to reformulate the current theoretical framework for optimization when BN is present.

© The Author(s), under exclusive license to Springer Nature Switzerland AG 2022
L. Huang, *Normalization Techniques in Deep Learning*, Synthesis Lectures on Computer Vision, https://doi.org/10.1007/978-3-031-14595-7_11

In contrast, normalizing-weights methods do not harm the theoretic analysis of DNNs, and can even attribute to boosting the theoretical results. For example, the Lipschitz constant w.r.t.a linear layer can be controlled/bounded during training by normalizing the weight with (approximate) orthogonality [7, 8], which is an important property for certified defense against adversarial attacks [9–11], and for theoretically analyzing DNN's generalization [12, 13]. However, normalizing weights is still not as effective as normalizing activations when it comes to improving training performance, leaving room for further development.

Applications Perspective: As mentioned previously, normalization methods can be used to 'edit' the statistical properties of layer activations, which has been exploited in CV tasks to match particular domain knowledges. However, we note that this mechanism is seldom used in NLP tasks. It would thus be interesting to investigate the correlation between the statistical properties of layer activations and the domain knowledge in NLP, and further improve the performance of the corresponding tasks. In addition, There exists an intriguing phenomenon that, while BN/GN work for the CV models, LN is more effective in NLP [14]. Intuitively, BN/GN should work well for NLP tasks, considering that the current state-of-the-art models for CV and NLP tend to be similar (e.g., they both use the convolutional operation and attention) and GN is simply a more general version of LN. It is thus important to further investigate whether or not BN/GN can be made to work well for NLP tasks, and, if not, why.

Another interesting observation is that normalization is not very common in deep reinforcement learning (DRL) [15]. Considering that certain DRL frameworks (e.g., actor-critic [16, 17]) are very similar to GANs, it should be possible to exploit normalization techniques to improve training in DRL, borrowing ideas from GANs (e.g., normalizing the weights in the discriminator [7, 8, 18]).

As the key components in DNNs, normalization techniques are links that connect the theory and application of deep learning. We thus believe that these techniques will continue to have a profound impact on the rapidly growing field of deep learning, and we hope that this book will aid readers in building a comprehensive landscape for their implementation.

References

1. Montufar, G. F., R. Pascanu, K. Cho, and Y. Bengio (2014). On the number of linear regions of deep neural networks. In *NeurIPS*.
2. Sun, R. (2019). Optimization for deep learning: theory and algorithms. arXiv preprint arXiv: 1912.08957.
3. Zhang, C., S. Bengio, M. Hardt, B. Recht, and O. Vinyals (2017). Understanding deep learning requires rethinking generalization. In *ICLR*.
4. Yang, G., J. Pennington, V. Rao, J. Sohl-Dickstein, and S. S. Schoenholz (2019). A mean field theory of batch normalization. In *ICLR*.
5. Xiong, H., L. Huang, M. Yu, L. Liu, F. Zhu, and L. Shao (2020). On the number of linear regions of convolutional neural networks. In *ICML*.

6. Lian, X. and J. Liu (2019). Revisit batch normalization: New understanding and refinement via composition optimization. In *AISTATS*.
7. Miyato, T., T. Kataoka, M. Koyama, and Y. Yoshida (2018). Spectral normalization for generative adversarial networks. In *ICLR*.
8. Huang, L., L. Liu, F. Zhu, D. Wan, Z. Yuan, B. Li, and L. Shao (2020). Controllable orthogonalization in training DNNs. In *CVPR*.
9. Tsuzuku, Y., I. Sato, and M. Sugiyama (2018). Lipschitz-margin training: Scalable certification of perturbation invariance for deep neural networks. In *NeurIPS*.
10. Anil, C., J. Lucas, and R. Grosse (2019). Sorting out lipschitz function approximation. In *ICLR*.
11. Qian, H. and M. N. Wegman (2019). L2-nonexpansive neural networks. In *ICLR*.
12. Bartlett, P. L., D. J. Foster, and M. J. Telgarsky (2017). Spectrally-normalized margin bounds for neural networks. In *NeurIPS*.
13. Neyshabur, B., S. Bhojanapalli, and N. Srebro (2018). A PAC-bayesian approach to spectrally-normalized margin bounds for neural networks. In *ICLR*.
14. Shen, S., Z. Yao, A. Gholami, M. W. Mahoney, and K. Keutzer (2020). Powernorm: Rethinking batch normalization in transformers. In *ICML*.
15. Bhatt, A., M. Argus, A. Amiranashvili, and T. Brox (2019). Crossnorm: Normalization for off-policy TD reinforcement learning. arXiv preprint arXiv:1902.05605.
16. Lillicrap, T. P., J. J. Hunt, A. Pritzel, N. Heess, T. Erez, Y. Tassa, D. Silver, and D. Wierstra (2016). Continuous control with deep reinforcement learning. In *ICLR*.
17. Mnih, V., A. P. Badia, M. Mirza, A. Graves, T. Lillicrap, T. Harley, D. Silver, and K. Kavukcuoglu (2016). Asynchronous methods for deep reinforcement learning. In *ICML*.
18. Brock, A., J. Donahue, and K. Simonyan (2019). Large scale GAN training for high fidelity natural image synthesis. In *ICLR*.

Appendix

A

A.1 Back-Propagation Through Eigenvalue Decomposition

Proposition: Considering the symmetric matrix $\Sigma \in R^{d \times d}$, its eigenvalue decomposition can be described as $\Sigma = D \Lambda D^T$, where $\Lambda = \text{diag}(\lambda_1, \ldots, \lambda_d)$ and D are the eigenvalues and eigenvectors of the covariance matrix Σ. We have $D^T D = I$. $\mathcal{L}$ is the loss function which depends on D and Λ. Therefore, $\mathcal{L}$ depends on Σ. Given $\frac{\partial \mathcal{L}}{\partial D}$ and $\frac{\partial \mathcal{L}}{\partial \Lambda}$, we have:

$$\frac{\partial \mathcal{L}}{\partial \Sigma} = D\{(K^T \odot (D^T \frac{\partial \mathcal{L}}{\partial D})) + (\frac{\partial \mathcal{L}}{\partial \Lambda})_{diag}\} D^T \tag{A.1}$$

where $K \in \mathbb{R}^{d \times d}$ is 0-diagonal and structured as $K_{ij} = \frac{1}{\lambda_i - \lambda_j}[i \neq j]$, the $\odot$ operator represents element-wise matrix multiplication, and $(\frac{\partial \mathcal{L}}{\partial \Lambda})_{diag}$ sets all off-diagonal elements of $\frac{\partial \mathcal{L}}{\partial \Lambda}$ to zero.

Proof The key idea of proof is based on the chain rule and perturbation theory. Based on the chain rule, we have

$$\frac{\partial \mathcal{L}}{\partial \Sigma_{ij}} = \sum_{k=1}^{d} \frac{\partial \mathcal{L}}{\partial \lambda_k} \frac{\partial \lambda_k}{\partial \Sigma_{ij}} + \sum_{n=1}^{d} \sum_{m=1}^{d} \frac{\partial \mathcal{L}}{\partial D_{nm}} \frac{\partial D_{nm}}{\partial \Sigma_{ij}} \tag{A.2}$$

where λ_k is the kth eigenvalue. The next step is to calculate $\frac{\partial \Lambda}{\partial \Sigma}$ and $\frac{\partial D}{\partial \Sigma}$, given the eigenvalue decomposition $\Sigma = D \Lambda D^T$, with $\Sigma \in \mathbb{R}^{d \times d}$, $\Lambda \in \mathbb{R}^{d \times d}$ diagonal and $D \in \mathbb{R}^{d \times d}$ orthogonal. The respective constrains are that: (1) the variation $\partial \Lambda$ is diagonal, like Λ; (2) ∂D satisfies the constraint that $D^T \partial D + \partial D^T D = 0$, which is derived by $D^T D = I$. Firstly, we try to derive $\frac{\partial \Lambda}{\partial \Sigma}$. We can get $\Lambda = D^T \Sigma D$. Taking the first variation of $\Lambda = D^T \Sigma D$, we have

© The Editor(s) (if applicable) and The Author(s), under exclusive license to Springer
Nature Switzerland AG 2022
L. Huang, *Normalization Techniques in Deep Learning*, Synthesis Lectures on Computer
Vision, https://doi.org/10.1007/978-3-031-14595-7

$$\partial \Lambda = \partial \boldsymbol{D}^T \Sigma \boldsymbol{D} + \boldsymbol{D}^T \partial \Sigma \boldsymbol{D} + \boldsymbol{D}^T \Sigma \partial \boldsymbol{D}. \tag{A.3}$$

Note that $\partial \Lambda$ is a diagonal matrix based on the constrain. By using $\Sigma \boldsymbol{D} = \boldsymbol{D}\Lambda$ and $\boldsymbol{D}^T \Sigma = \Lambda \boldsymbol{D}^T$, we can get

$$\partial \Lambda = \partial \boldsymbol{D}^T \boldsymbol{D}\Lambda + \boldsymbol{D}^T \partial \Sigma \boldsymbol{D} + \Lambda \boldsymbol{D}^T \partial \boldsymbol{D}. \tag{A.4}$$

Let $\boldsymbol{A} = \boldsymbol{D}^T \partial \boldsymbol{D}$, and based on constrain that $\boldsymbol{D}^T \partial \boldsymbol{D} + \partial \boldsymbol{D}^T \boldsymbol{D} = 0$, we have $\boldsymbol{A} + \boldsymbol{A}^T = 0$, which means that $\boldsymbol{A}$ is antisymmetric. Therefore, the diagonal elements of $\boldsymbol{A}$ are zeros. Whereas $\boldsymbol{A}\Lambda$ and $\Lambda \boldsymbol{A}$ are both zero diagonal, and since $\partial \Lambda$ is diagonal, we can get

$$\partial \Lambda = (\boldsymbol{D}^T \partial \Sigma \boldsymbol{D})_{diag} \tag{A.5}$$

and therefore we have

$$\frac{\partial \lambda_k}{\partial \Sigma_{ij}} = \boldsymbol{D}_{ik} \boldsymbol{D}_{jk} \tag{A.6}$$

Secondly, we derive $\frac{\partial \boldsymbol{D}}{\partial \Sigma}$. Denoting $\mathbf{V} = \boldsymbol{D}^T \partial \Sigma \boldsymbol{D} - \partial \Lambda$, we can get

$$\boldsymbol{A}\Lambda - \Lambda \boldsymbol{A} = \mathbf{V} \Rightarrow \begin{cases} A_{ij}\lambda_j - A_{ij}\lambda_i = V_{ij} & i \neq j \\ A_{ij} = 0 & i = j \end{cases} \tag{A.7}$$

so that we have $\boldsymbol{A} = \boldsymbol{K}^T \odot \mathbf{V}$, where $\odot$ is element-wise multiply operation and K has elements:

$$K_{ij} = \begin{cases} \frac{1}{\lambda_i - \lambda_j} & i \neq j \\ 0 & i = j. \end{cases}$$

We thus have $\boldsymbol{A} = \boldsymbol{K}^T \odot \mathbf{V} = \boldsymbol{K}^T \odot (\boldsymbol{D}^T \partial \Sigma \boldsymbol{D}) - \boldsymbol{K}^T \odot \partial \Lambda = \boldsymbol{K}^T \odot (\boldsymbol{D}^T \partial \Sigma \boldsymbol{D})$. Therefore, we can get

$$\partial \boldsymbol{D} = \boldsymbol{D}(\boldsymbol{K}^T \odot (\boldsymbol{D}^T \partial \Sigma \boldsymbol{D})) \tag{A.8}$$

and that is

$$\frac{\partial \boldsymbol{D}_{nm}}{\partial \Sigma_{ij}} = \sum_{s \neq m} \frac{\boldsymbol{D}_{ns} \boldsymbol{D}_{is} \boldsymbol{D}_{jm}}{\lambda_m - \lambda_s}. \tag{A.9}$$

Based on Eqs. (A.2), (A.6) and (A.9), we have:

$$\frac{\partial \mathcal{L}}{\partial \Sigma_{ij}} = \sum_{k=1}^{d} \frac{\partial \mathcal{L}}{\partial \lambda_k} D_{ik} D_{jk} + \sum_{n=1}^{d} \sum_{m=1}^{d} \frac{\partial \mathcal{L}}{\partial D_{nm}} \sum_{s \neq m} \frac{D_{ns} D_{is} D_{jm}}{\lambda_m - \lambda_s} \tag{A.10}$$

formed in matrix by

$$\frac{\partial \mathcal{L}}{\partial \Sigma} = D(\frac{\partial \mathcal{L}}{\partial \Lambda})_{diag} D^T + D\{(K^T \odot (D^T \frac{\partial \mathcal{L}}{\partial D}))\} D^T. \tag{A.11}$$

Thus, we can get Eq. (A.1). $\qquad\qquad\qquad\qquad\qquad\qquad\qquad\qquad\qquad\qquad\square$

A.2 Derivation of Constraint Number of Normalization Methods

In Sect. 9.4 of the book, we define the constraint number of a normalization operation, and summarize the constraint number of different normalization methods in Table 9.1 of the book. Here, we provide the details for deriving the constraint number of batch whitening (BW), group normalization (GN) [1] and group whitening (GW), for the mini-batch input $\mathbf{X} \in \mathbb{R}^{d \times m}$.

Constraint number of BW. BW [2] ensures that the normalized output is centered and whitened, which has the constraints $\Upsilon_{\phi_{BW}}(\widehat{\mathbf{X}})$ as:

$$\widehat{\mathbf{X}}\mathbf{1} = \mathbf{0}_d, \quad and \tag{A.12}$$

$$\widehat{\mathbf{X}}\widehat{\mathbf{X}}^T - m\mathbf{I} = \mathbf{0}_{d \times d}, \tag{A.13}$$

where $\mathbf{0}_d$ is a d-dimensional column vector of all zeros, and $\mathbf{0}_{d \times d}$ is a $d \times d$ matrix of all zeros. Note that there are d independent equations in the system of equations $\widehat{\mathbf{X}}\mathbf{1} = \mathbf{0}_d$. Let's denote $\mathbf{M} = \widehat{\mathbf{X}}\widehat{\mathbf{X}}^T - m\mathbf{I}$. We have $\mathbf{M}^T = \mathbf{M}$, and thus $\mathbf{M}$ is a symmetric matrix. Therefore, there are $d(d+1)/2$ independent equations in the system of equations $\widehat{\mathbf{X}}\widehat{\mathbf{X}}^T - m\mathbf{I} = \mathbf{0}_{d \times d}$. We thus have $d(d+1)/2 + d$ independent equations in $\Upsilon_{\phi_{BW}}(\widehat{\mathbf{X}})$, and the constraint number of BW is $d(d+3)/2$.

Constraint number of GN. Given a sample $\mathbf{x} \in \mathbb{R}^d$, GN divides the neurons into groups: $\mathbf{Z} = \Pi(\mathbf{x}) \in \mathbb{R}^{g \times c}$, where g is the group number and $d = gc$. The standardization operation is then performed on $\mathbf{Z}$ as:

$$\widehat{\mathbf{Z}} = \Lambda_g^{-\frac{1}{2}}(\mathbf{Z} - \mu_g \mathbf{1}^T), \tag{A.14}$$

where, $\mu_g = \frac{1}{c}\mathbf{Z}\mathbf{1}$ and $\Lambda_g = \text{diag}(\lambda_1^2, \ldots, \lambda_g^2) + \epsilon \mathbf{I}$. This ensures that the normalized output $\widehat{\mathbf{Z}}$ for each sample has the constraints:

$$\sum_{j=1}^{c} \widehat{\mathbf{Z}}_{ij} = 0 \; and \; \sum_{j=1}^{c} \widehat{\mathbf{Z}}_{ij}^2 = c, \; for \; i = 1, \ldots, g. \tag{A.15}$$

In the system of Eq. (A.15), the number of independent equations is $2g$. Therefore, the constraint number of GN is $2dm$, when given m samples.

Constraint number of GW. Given a sample $\mathbf{x} \in \mathbb{R}^d$, GW performs normalization as:

$$Group\ division : \mathbf{X}_G = \Pi(\mathbf{x}; g) \in \mathbb{R}^{g \times c}, \tag{A.16}$$

$$Whitening : \widehat{\mathbf{X}}_G = \Sigma_g^{-\frac{1}{2}}(\mathbf{X}_G - \mu_g \mathbf{1}^T), \tag{A.17}$$

$$Inverse\ group\ division : \hat{\mathbf{x}} = \Pi^{-1}(\widehat{\mathbf{X}}_G) \in \mathbb{R}^d. \tag{A.18}$$

The normalization operation ensures that $\widehat{\mathbf{X}}_G \in \mathbb{R}^{g \times c}$ has the following constraints:

$$\widehat{\mathbf{X}}_G \mathbf{1} = \mathbf{0}, \quad and \tag{A.19}$$

$$\widehat{\mathbf{X}}_G \widehat{\mathbf{X}}_G^T - c\mathbf{I} = \mathbf{0}. \tag{A.20}$$

Following the analysis for BW, the number of independent equations is $g(g+3)/2$ from Eqs. (A.19) and (A.20). Therefore, the constraint number of GW is $mg(g+3)/2$, when given m samples.

A.3 Proofs of Theorems

Theorem 1 *Given a rectifier neural network with nonlinearity $\phi(\alpha \mathbf{x}) = \alpha\phi(\mathbf{x})$ ($\alpha > 0$), if the weight in each layer is scaled by $\widehat{\mathbf{W}}_l = \alpha_l \mathbf{W}_l$ ($l = 1, \ldots, L$ and $\alpha_l > 0$), we have the scaled layer input: $\widehat{\mathbf{x}}_l = \left(\prod_{i=1}^{l} \alpha_i\right) \mathbf{x}_l$. Assuming that $\frac{\partial \mathcal{L}}{\partial \mathbf{h}_L} = \mu \frac{\partial \mathcal{L}}{\partial \mathbf{h}_L}$, we have the output-gradient: $\frac{\partial \mathcal{L}}{\partial \widehat{\mathbf{h}}_l} = \mu \left(\prod_{i=l+1}^{L} \alpha_i\right) \frac{\partial \mathcal{L}}{\partial \mathbf{h}_l}$, and weight-gradient: $\frac{\partial \mathcal{L}}{\partial \widehat{\mathbf{W}}_l} = \left(\mu \prod_{i=1, i \neq l}^{L} \alpha_i\right) \frac{\partial \mathcal{L}}{\partial \mathbf{W}_l}$, for all $l = 1, \ldots, L$.*

Proof (1) We first demonstrate that the scaled layer input $\widehat{\mathbf{x}}_l = \left(\prod_{i=1}^{l} \alpha_i\right) \mathbf{x}_l$ ($l = 1, \ldots, L$), using mathematical induction. It is easy to validate that $\widehat{\mathbf{h}}_1 = \alpha_1 \mathbf{h}_1$ and $\widehat{\mathbf{x}}_1 = \alpha_1 \mathbf{x}_1$. We assume that $\widehat{\mathbf{h}}_t = \left(\prod_{i=1}^{t} \alpha_i\right) \mathbf{h}_t$ and $\widehat{\mathbf{x}}_t = \left(\prod_{i=1}^{t} \alpha_i\right) \mathbf{x}_t$ hold, for $t = 1, \ldots, l$. When $t = l+1$, we have

$$\widehat{\mathbf{h}}_{l+1} = \widehat{\mathbf{W}}_{l+1} \widehat{\mathbf{x}}_{l+1} = \alpha_{l+1} \mathbf{W}_{l+1} \left(\prod_{i=1}^{l} \alpha_i\right) \mathbf{x}_l = \left(\prod_{i=1}^{l+1} \alpha_i\right) \mathbf{h}_{l+1}. \tag{A.21}$$

We thus have

$$\hat{\mathbf{x}}_{l+1} = \phi(\hat{\mathbf{h}}_{l+1}) = \phi\left(\left(\prod_{i=1}^{l+1} \alpha_i\right) \mathbf{h}_{l+1}\right) = \left(\prod_{i=1}^{l+1} \alpha_i\right) \phi(\mathbf{h}_{l+1}) = \left(\prod_{i=1}^{l+1} \alpha_i\right) \mathbf{x}_{l+1}. \text{(A.22)}$$

By induction, we have $\hat{\mathbf{x}}_l = \left(\prod_{i=1}^{l} \alpha_i\right) \mathbf{x}_l$, for $l = 1, \ldots, L$. We also have $\hat{\mathbf{h}}_l = \left(\prod_{i=1}^{l} \alpha_i\right) \mathbf{h}_l$ for $l = 1, \ldots, L$.

(2) We then demonstrate that the scaled output-gradient $\frac{\partial \mathcal{L}}{\partial \hat{\mathbf{h}}_l} = \mu \left(\prod_{i=l+1}^{K} \alpha_i\right) \frac{\partial \mathcal{L}}{\partial \mathbf{h}_l}$ for $l = 1, \ldots, L$. We also provide this using mathematical induction. Based on back-propagation, we have

$$\frac{\partial \mathcal{L}}{\partial \mathbf{x}_{l-1}} = \frac{\partial \mathcal{L}}{\partial \mathbf{h}_l} \mathbf{W}_l, \quad \frac{\partial \mathcal{L}}{\partial \mathbf{h}_{l-1}} = \frac{\partial \mathcal{L}}{\partial \mathbf{x}_{l-1}} \frac{\partial \mathbf{x}_{l-1}}{\partial \mathbf{h}_{l-1}}, \tag{A.23}$$

and

$$\frac{\partial \hat{\mathbf{x}}_{l-1}}{\partial \hat{\mathbf{h}}_{l-1}} = \frac{\partial \left(\prod_{i=1}^{l-1} \alpha_i\right) \mathbf{x}_{l-1}}{\partial \left(\prod_{i=1}^{l-1} \alpha_i\right) \mathbf{h}_{l-1}} = \frac{\left(\prod_{i=1}^{l-1} \alpha_i\right) \partial \mathbf{x}_{l-1}}{\left(\prod_{i=1}^{l-1} \alpha_i\right) \partial \mathbf{h}_{l-1}} = \frac{\partial \mathbf{x}_{l-1}}{\partial \mathbf{h}_{l-1}}, \quad l = 2, \ldots, L. \tag{A.24}$$

Based on the assumption that $\frac{\partial \mathcal{L}}{\partial \hat{\mathbf{h}}_L} = \mu \frac{\partial \mathcal{L}}{\partial \mathbf{h}_L}$, we have $\frac{\partial \mathcal{L}}{\partial \hat{\mathbf{h}}_L} = \mu \left(\prod_{i=K+1}^{K} \alpha_i\right) \frac{\partial \mathcal{L}}{\partial \mathbf{h}_L}$ [1].

We assume that $\frac{\partial \mathcal{L}}{\partial \hat{\mathbf{h}}_t} = \mu \left(\prod_{i=t+1}^{L} \alpha_i\right) \frac{\partial \mathcal{L}}{\partial \mathbf{h}_t}$ holds, for $t = L, \ldots l$. When $t = l - 1$, we have

$$\frac{\partial \mathcal{L}}{\partial \hat{\mathbf{x}}_{l-1}} = \frac{\partial \mathcal{L}}{\partial \hat{\mathbf{h}}_l} \hat{\mathbf{W}}_l = \mu \left(\prod_{i=l+1}^{L} \alpha_i\right) \frac{\partial \mathcal{L}}{\partial \mathbf{h}_l} \alpha_l \mathbf{W}_l = \mu \left(\prod_{i=l}^{L} \alpha_i\right) \frac{\partial \mathcal{L}}{\partial \mathbf{x}_{l-1}}. \tag{A.25}$$

We also have

$$\frac{\partial \mathcal{L}}{\partial \hat{\mathbf{h}}_{l-1}} = \frac{\partial \mathcal{L}}{\partial \hat{\mathbf{x}}_{l-1}} \cdot \frac{\partial \hat{\mathbf{x}}_{l-1}}{\partial \hat{\mathbf{h}}_{l-1}} = \mu \left(\prod_{i=l}^{L} \alpha_i\right) \frac{\partial \mathcal{L}}{\partial \mathbf{x}_{l-1}} \cdot \frac{\partial \mathbf{x}_{l-1}}{\partial \mathbf{h}_{l-1}} = \mu \left(\prod_{i=l}^{L} \alpha_i\right) \frac{\partial \mathcal{L}}{\partial \mathbf{h}_{l-1}}. \tag{A.26}$$

By induction, we thus have $\frac{\partial \mathcal{L}}{\partial \hat{\mathbf{h}}_l} = \mu \left(\prod_{i=l+1}^{L} \alpha_i\right) \frac{\partial \mathcal{L}}{\partial \mathbf{h}_l}$, for $l = 1, \ldots, L$.

---

[1] We denote $\prod_{i=a}^{b} \alpha_i = 1$ if $a > b$.

(3) Based on $\frac{\partial \mathcal{L}}{\partial \mathbf{W}_l} = \frac{\partial \mathcal{L}}{\partial \mathbf{h}_l}^T \mathbf{x}_{l-1}^T$, $\widehat{\mathbf{x}}_l = \left(\prod_{i=1}^{l} \alpha_i \right) \mathbf{x}_l$ and $\frac{\partial \mathcal{L}}{\partial \widehat{\mathbf{h}}_l} = \mu \left(\prod_{i=l+1}^{L} \alpha_i \right) \frac{\partial \mathcal{L}}{\partial \widehat{\mathbf{h}}_l}$, it is easy

to prove that $\frac{\partial \mathcal{L}}{\partial \widehat{\mathbf{W}}_l} = \left(\mu \prod_{i=1, i \neq l}^{L} \alpha_i \right) \frac{\partial \mathcal{L}}{\partial \mathbf{W}_l}$ for $l = 1, \dots, L$. $\qquad\square$

Theorem 2 *Under the same condition as Theorem 1, for the normalized network with* $\mathbf{h}_l = \mathbf{W}_l \mathbf{x}_{l-1}$ *and* $\mathbf{s}_l = BN(\mathbf{h}_l)$, *we have:* $\hat{\mathbf{x}}_l = \mathbf{x}_l$, $\frac{\partial \mathcal{L}}{\partial \widehat{\mathbf{h}}_l} = \frac{1}{\alpha_l} \frac{\partial \mathcal{L}}{\partial \mathbf{h}_l}$, $\frac{\partial \mathcal{L}}{\partial \widehat{\mathbf{W}}_l} = \frac{1}{\alpha_l} \frac{\partial \mathcal{L}}{\partial \mathbf{W}_l}$, *for all* $l = 1, \dots, L$.

Proof (1) Following the proof in Theorem 1, by mathematical induction, it is easy to demonstrate that $\hat{\mathbf{h}}_l = \alpha_l \mathbf{h}_l$, $\widehat{\mathbf{s}}_l = \mathbf{s}_l$ and $\hat{\mathbf{x}}_l = \mathbf{x}_l$, for all $l = 1, \dots, L$.

(2) We also use mathematical induction to demonstrate $\frac{\partial \mathcal{L}}{\partial \widehat{\mathbf{h}}_l} = \frac{1}{\alpha_l} \frac{\partial \mathcal{L}}{\partial \mathbf{h}_l}$ for all $l = 1, \dots, L$. We first show the formulation of the gradient back-propagating through each neuron of the BN layer as:

$$\frac{\partial \mathcal{L}}{\partial h} = \frac{1}{\sigma} \left(\frac{\partial \mathcal{L}}{\partial s} - \mathbb{E}_{\mathcal{B}} \left(\frac{\partial \mathcal{L}}{\partial s} \right) - \mathbb{E}_{\mathcal{B}} \left(\frac{\partial \mathcal{L}}{\partial s} s \right) s \right), \qquad (A.27)$$

where σ is the standard deviation and $\mathbb{E}_{\mathcal{B}}$ denotes the expectation over mini-batch examples. We have $\hat{\sigma}_L = \alpha_L \sigma_L$ based on $\hat{\mathbf{h}}_L = \alpha_L \mathbf{h}_L$. Since $\widehat{\mathbf{s}}_L = \mathbf{s}_L$, we have $\frac{\partial \mathcal{L}}{\partial \widehat{\mathbf{s}}_L} = \frac{\partial \mathcal{L}}{\partial \mathbf{s}_L}$. Therefore, we have $\frac{\partial \mathcal{L}}{\partial \widehat{\mathbf{h}}_L} = \frac{\sigma_L}{\hat{\sigma}_L} \frac{\partial \mathcal{L}}{\partial \mathbf{h}_L} = \frac{1}{\alpha_L} \frac{\partial \mathcal{L}}{\partial \mathbf{h}_L}$ from Eq. A.27.

Assume that $\frac{\partial \mathcal{L}}{\partial \widehat{\mathbf{h}}_t} = \frac{1}{\alpha_t} \frac{\partial \mathcal{L}}{\partial \mathbf{h}_t}$ for $t = L, \dots, l+1$. When $t = l$, we have:

$$\frac{\partial \mathcal{L}}{\partial \hat{\mathbf{x}}_l} = \frac{\partial \mathcal{L}}{\partial \hat{\mathbf{h}}_{l+1}} \widehat{\mathbf{W}}_{l+1} = \frac{1}{\alpha_{l+1}} \frac{\partial \mathcal{L}}{\partial \mathbf{h}_{l+1}} \alpha_{l+1} \mathbf{W}_{l+1} = \frac{\partial \mathcal{L}}{\partial \mathbf{x}_l}. \qquad (A.28)$$

Following the proof for Theorem 1, it is easy to get $\frac{\partial \mathcal{L}}{\partial \widehat{\mathbf{s}}_l} = \frac{\partial \mathcal{L}}{\partial \mathbf{s}_l}$. Based on $\frac{\partial \mathcal{L}}{\partial \widehat{\mathbf{s}}_l} = \frac{\partial \mathcal{L}}{\partial \mathbf{s}_l}$ and $\widehat{\mathbf{s}}_l = \mathbf{s}_l$, we have $\frac{\partial \mathcal{L}}{\partial \widehat{\mathbf{h}}_l} = \frac{\sigma_l}{\hat{\sigma}_l} \frac{\partial \mathcal{L}}{\partial \mathbf{h}_l} = \frac{1}{\alpha_l} \frac{\partial \mathcal{L}}{\partial \mathbf{h}_l}$ from Eq. (A.27).

By induction, we have $\frac{\partial \mathcal{L}}{\partial \widehat{\mathbf{h}}_l} = \frac{1}{\alpha_l} \frac{\partial \mathcal{L}}{\partial \mathbf{h}_l}$, for all $l = 1, \dots, L$.

(3) Based on $\frac{\partial \mathcal{L}}{\partial \mathbf{W}_l} = \frac{\partial \mathcal{L}}{\partial \mathbf{h}_l}^T \mathbf{x}_{l-1}^T$, $\hat{\mathbf{x}}_l = \mathbf{x}_l$ and $\frac{\partial \mathcal{L}}{\partial \widehat{\mathbf{h}}_l} = \frac{1}{\alpha_l} \frac{\partial \mathcal{L}}{\partial \mathbf{h}_l}$, we have that $\frac{\partial \mathcal{L}}{\partial \widehat{\mathbf{W}}_l} = \frac{1}{\alpha_l} \frac{\partial \mathcal{L}}{\partial \mathbf{W}_l}$, for all $l = 1, \dots, L$. $\qquad\square$

References

1. Wu, Y. and K. He (2018). Group normalization. In *ECCV*.
2. Huang, L., D. Yang, B. Lang, and J. Deng (2018). Decorrelated batch normalization. In *CVPR*.